science
works 2
David Blaker and
Edgar Mitchell
CENGAGE
Learning™
Australia • Brazil • Japan • Korea • Mexico • Singapore • Spain • United Kingdom • United States

Science Works 2
1st Edition
David Blaker
Edgar Mitchell

Cover and text design: Helen Andrewes
Illustrations□: Richard Gunther
Typeset by Helen Andrewes

Any URLs contained in this publication were checked for currency during the production process. Note, however, that the publisher cannot vouch for the ongoing currency of URLs.

For product information and technology assistance,
in Australia call 1300 790 853;
in New Zealand call 0800 449 725

For permission to use material from this text or product, please email aust.permissions@cengage.com

National Library of Australia Cataloguing-in-Publication Data
National Library of New Zealand Cataloguing-in-Publication Data

Blaker, David (David Neville)
Science works. 2 / David Blaker and Edgar Mitchell.
ISBN 978-017095-013-8
1. Science—Problems, exercises, etc.—Juvenile literature.
2. Science—Problems, exercises, etc. I. Mitchell, Edgar D.
II. Title.
507.6—dc 22

Cengage Learning Australia
Level 7, 80 Dorcas Street
South Melbourne, Victoria Australia 3205

Cengage Learning New Zealand
Unit 4B Rosedale Office Park
331 Rosedale Road, Albany, North Shore 0632, NZ

For learning solutions, visit cengage.com.au

Printed in Australia by Ligare Pty Limited.
2 3 4 5 6 7 8 20 19 18 17 16

CONTENTS

KITCHEN action

In this chapter we will:

- Learn about chemicals in the kitchen
- Learn about cooking with chemical reactions
- Learn what makes flavours
- Learn how detergents work

A KITCHEN is full of chemicals. Some of these have been made by plants. Some have been made in factories.

Some examples of foods with natural chemicals in them:

- **sugar** is pure sucrose
- **salt** is pure sodium chloride
- **vinegar** contains acetic acid
- **oranges** contain citric acid.

Bread, buns, flour and pasta are all made from the ground-up grains of wheat plants. Starch is the main chemical substance in wheat grains and other kinds of seeds. Natural whole foods like tomato or pumpkin have hundreds of different chemical substances in them.

What do we mean by 'natural'? Usually this means a food with not many added chemicals. Let's look at some examples, and you can decide where to draw the line between natural and artificial.

- Raw, uncooked, unprocessed, like fresh fruit.
- Uncooked, partly processed, like milk or cheese.
- Cooked and partly processed, like bread.
- Cooked and highly processed, like *Cheezles.*

Here are some of the ingredients listed on a *Cheezles* packet:

Ingredients: corn, vegetable oil, antioxidant 319, cheese flavour, tapioca starch, flavour enhancers 621 and 635, sugar, food acids 270 and 330, emulsifier 471, colour 160e.

What does cooking do to food?

Some foods taste good raw, like apple or avocado. As for others: raw egg and raw meat won't kill you, but the taste is not great. Cooking involves chemical reactions. You can be sure if something is a chemical change, because it can't be reversed. You can't uncook an egg!

When you make toast, some of the starch changes to caramel, which gives it a toasty flavour.

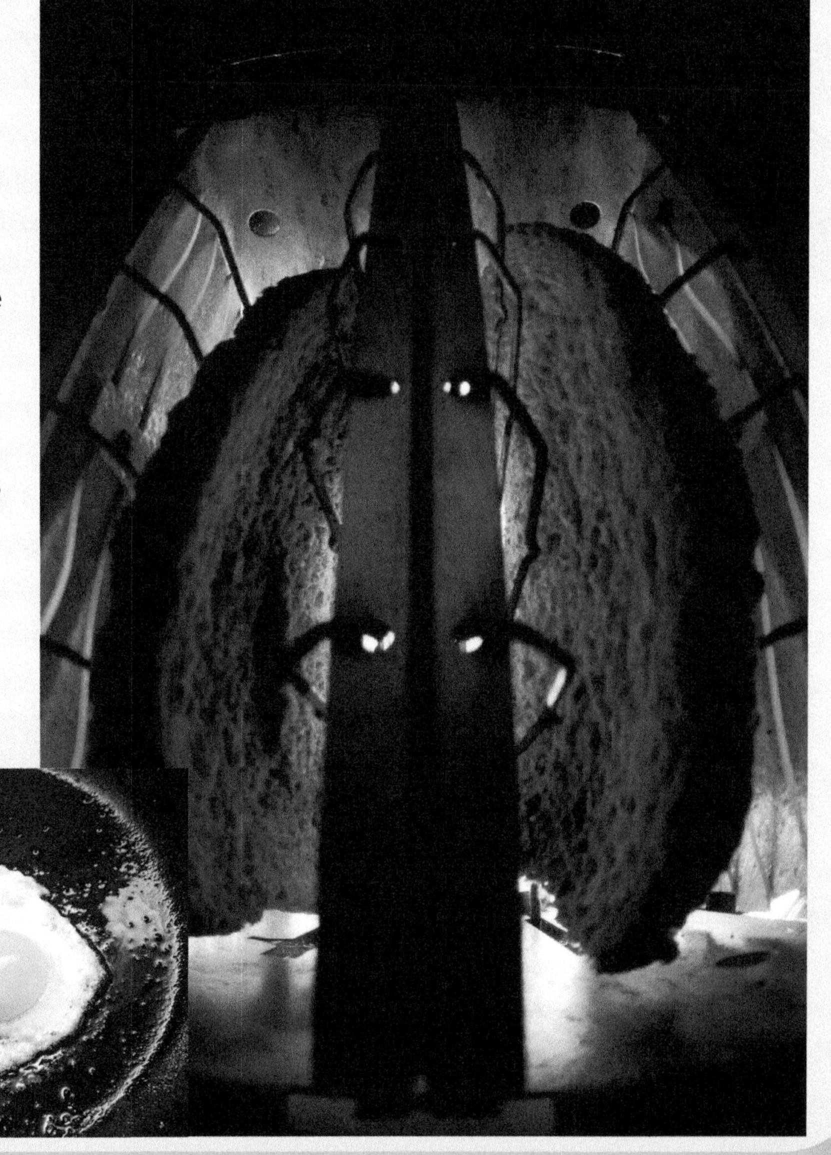

And meat?

Meat is about 30% protein – the other 70% is fat and water. Proteins are natural substances like long chains. Cooking makes these chains fold into new shapes.

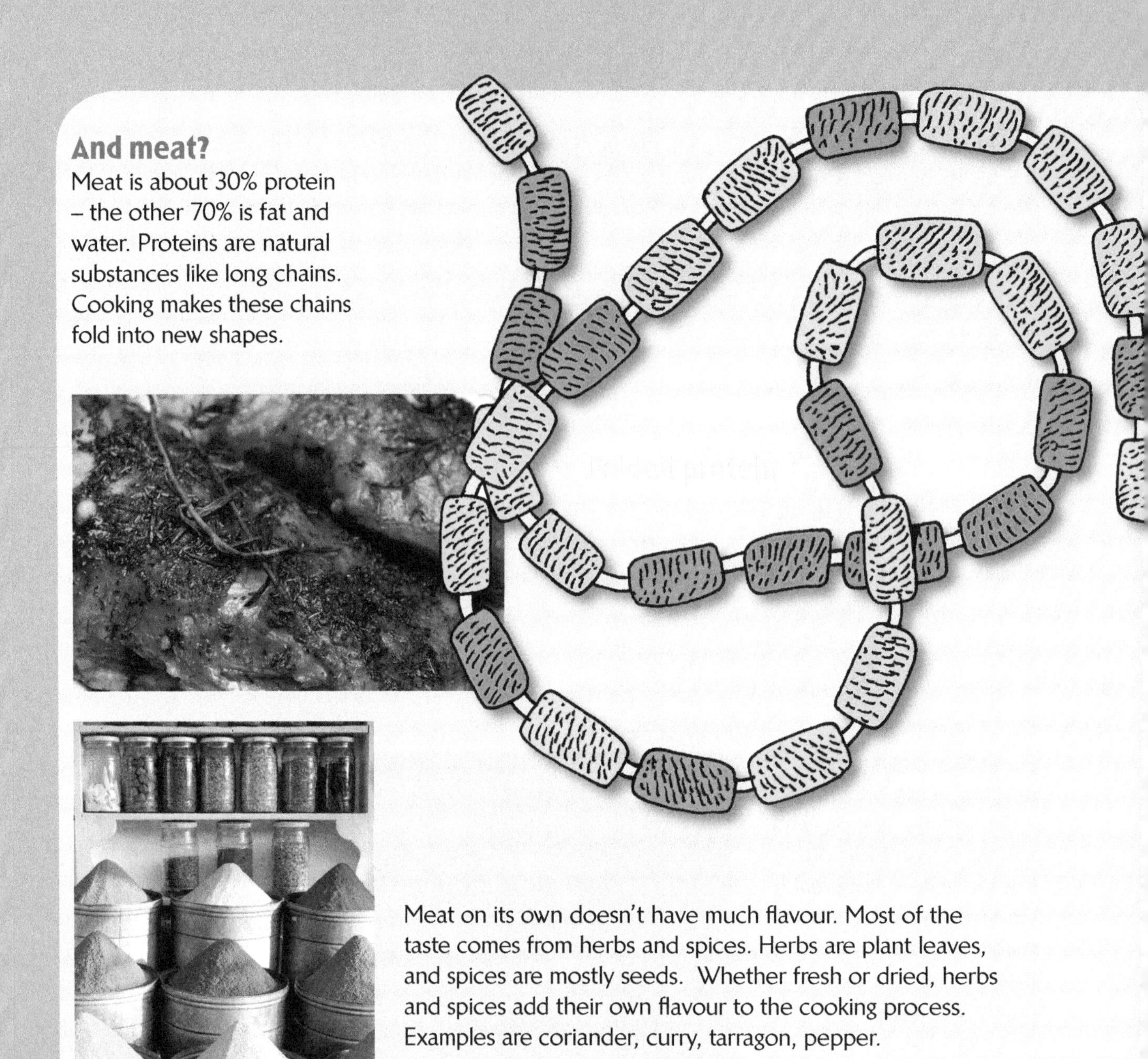

Meat on its own doesn't have much flavour. Most of the taste comes from herbs and spices. Herbs are plant leaves, and spices are mostly seeds. Whether fresh or dried, herbs and spices add their own flavour to the cooking process. Examples are coriander, curry, tarragon, pepper.

Question:

What causes pancakes and ordinary cakes to be full of little holes?

Answer:

Baking powder in the cake mix. During cooking the baking powder gives off bubbles of CO_2 gas. (CO_2 is a natural gas also known as carbon dioxide.) These gas bubbles stay trapped and give the mixture a fluffy texture.

How does a detergent work?

As everybody knows, oil and water will not mix, which is another way of saying they don't dissolve in each other. This is why you can't wash greasy plates or spoons in plain water.

Detergents work in a clever way. Each detergent particle has a long narrow shape. One end can mix in water, but not in fat. The other end mixes in fat, but not in water. Scientists actually call these ends 'water-loving' and 'water-hating'! Their split personality means that a detergent can mix in water and fat at the same time.

Detergent at work

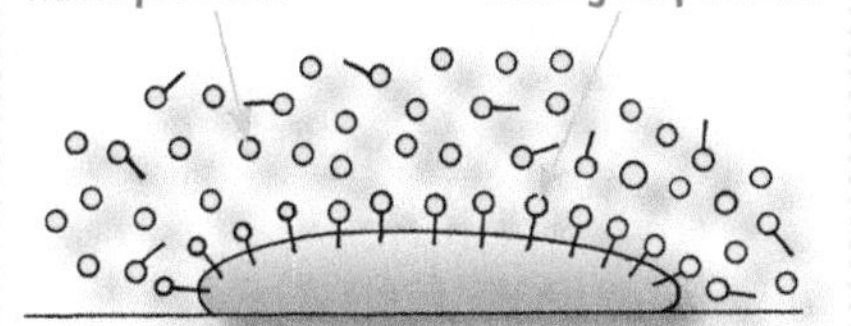

BEFORE Grease sticks to the surface.

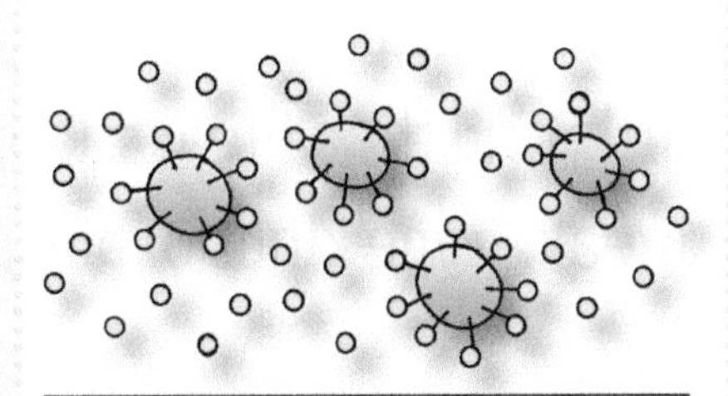

AFTER One end of each detergent particle is attracted to water, which lifts small drops of grease off the surface.

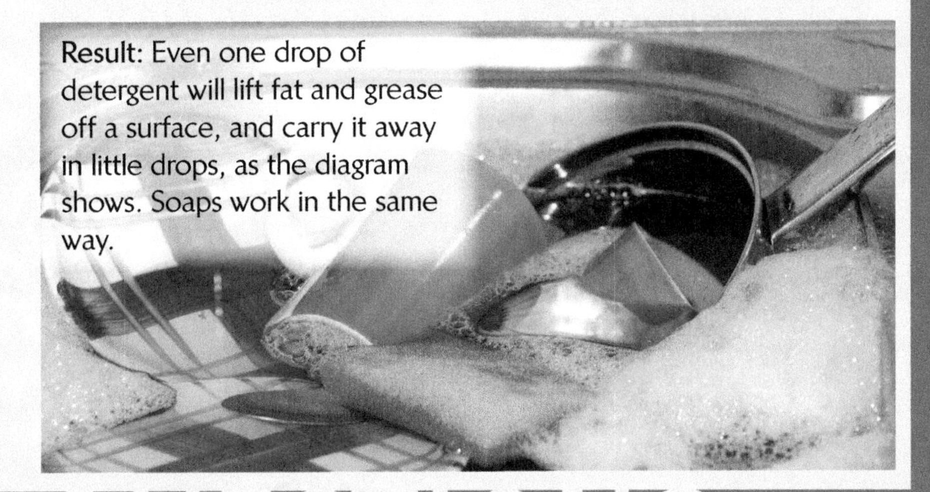

Result: Even one drop of detergent will lift fat and grease off a surface, and carry it away in little drops, as the diagram shows. Soaps work in the same way.

Hands-on

Here is a recipe for pancakes.

You will need:

- 1 cup of wholemeal flour (NOT self-raising flour)
- 3/4 cup of milk
- 1/8 teaspoon salt
- 1 egg
- 1/4 teaspoon baking powder

Mix them all thoroughly in a bowl and leave it to stand for a while. If the mixture looks too thick, stir in a little water. Put a label marked 'A' on the mixing bowl.

Now make another mixture in bowl 'B', exactly the same, but with NO baking powder.

Get two pans, heat them on two hot plates on the same stove, and put a small amount of butter in each. When the pans are hot, put a scoop of mix A into one pan, and a scoop of mix B into the other. When the bubbles start to burst, turn each 'cake' over and cook the other side.

One of these recipes will give fluffy pancakes, the other will be more like hard biscuits. Explain why this happens.

QUESTIONS

1. Name the chemical substance in sugar.
2. Name the chemical substance in salt.
3. Name the chemical substance in vinegar.
4. Name two chemical substances in orange juice.
4. Explain why toasting changes the taste of bread.
5. Name four different herbs and spices.
6. What is the difference between a herb and a spice?
7. Describe what baking powder does to cakes. Explain how it works.
8. Use words and drawings to explain how detergents work.

Extension:

Go to a supermarket (or the Internet) and make a list of 10 or more herbs and spices available. Now use the Internet or library to find out what kind of plant, and what part of the plant, each one comes from.

BOUNCING back

DIFFERENT MATERIALS have different features. Diamonds are hard. Steel is strong. Glass bends a little and then breaks. Rubber bends but does not break. Some kinds of plastic are light and strong.

What do we mean by 'plastic'?

The word has two meanings. First, it means substances that are used to make pens, rulers, toothbrush handles, supermarket bags. But the word 'plastic' has another meaning as well. 'Plastic' is a word we can use to describe anything that can be bent into a new shape, and stays that way. So even aluminium foil is 'plastic'.

'Elastic' means rubber bands, right?

Actually it means more than that. The word 'elastic' can be used for anything that is flexible and springs back into shape when you let go again. Rubber is very elastic. Wood is partly elastic

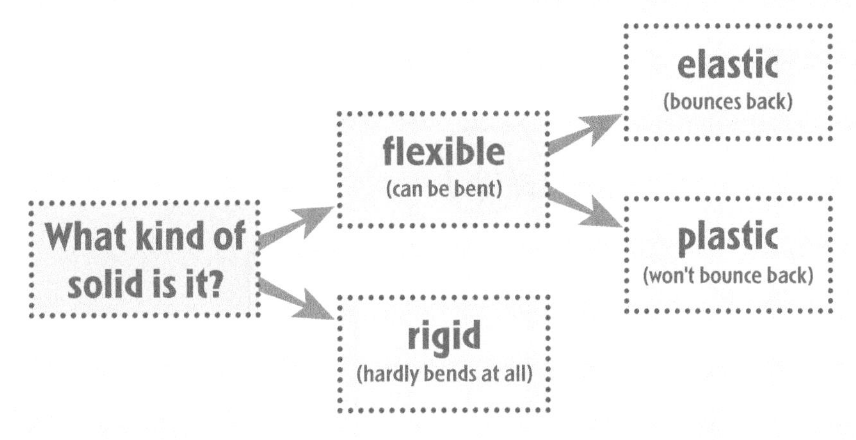

There are many kinds of substances that we know as 'plastic'. From bicycle helmets and flotation devices to mouth guards and other protective sporting gear, plastics help to keep active people safe and healthy.

Hands-on

The aim of this activity is to find which ball is bounciest.

Get several kinds of ball, such as a golf ball, tennis ball and basketball. You are going to drop them one at a time from a height (two metres is good) and find how high each one bounces.

The first thing is to make sure your tests are fair. Write down a list of all the things that should be kept the same when you do every bounce test.

Second, decide who does the different tasks in your group. One to drop the ball, one to mark and measure the bounce, one to write down the results in a neat table.

Third, it is a good idea to do each test three times, in case one result is completely wrong.

Finally, use a computer to draw a bar graph of your average bounce results for each ball.

Polymers

When you press on a rubber ball, or pull a rubber band and let go, they bounce back to their original shape — in other words they are elastic. To understand what causes this let's look at the polymers they are made of. A polymer is a special kind of chemical substance which has long chains of the smaller molecules, like a chain made out of paper clips.

Polymer chains can be easily bent, but want to pull back to the original shape again. Just like rubber — which has long polymers in it. Objects that are made out of polymers act differently depending on how the molecules are connected. Some feel rubbery, some are sticky, some are gooey, and some are hard and tough. Some examples of polymers are polythene, polystyrene and polycarbonate.

Hands-on

Collect items from around the house, and try to bend each one, then describe in your own words what they do. Use these words: flexible, not flexible, elastic, not elastic, plastic, not plastic.

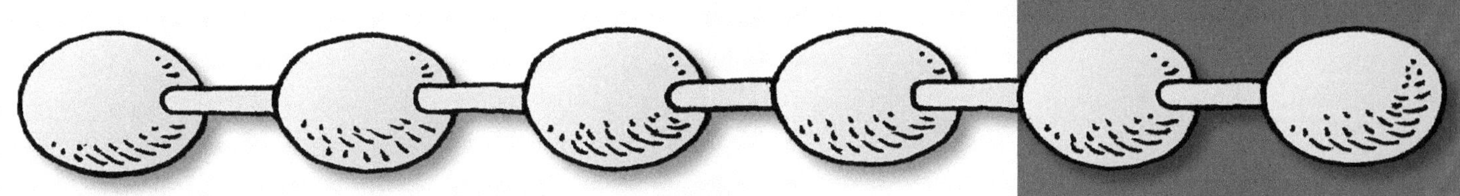

'Poly' means 'many'. 'Mer' means 'parts'. OK?

QUESTIONS

1. Say what the word 'polymer' means.
2. Say what the word 'plastic' means.
3. Say what the word 'elastic' means.
4. Say what the word 'flexible' means.
5. Explain in 15 to 30 words the difference between a material being 'plastic', and another material being 'elastic'.
6. Why are bicycle frames generally made out of metal? Give two reasons.
7. Describe the best material for making each of these things: crash helmet, mouth guard, trampoline mat. You don't have to name the actual material, but use some of these words in your answer: plastic, not plastic, elastic, not elastic.
8. You want to compare how high a golf ball bounces compared to a tennis ball. The test has to be a fair one. List four things you should do to make sure that each test is fair and each ball is tested in the same conditions.

Extension:

Find out how natural rubber is made, and from what. And what synthetic rubber is made from? What's special about polycarbonate?

COOKING, BURNING, reacting

In this chapter we will:

- Learn what happens to chemicals when they react
- Learn that some reactions are slow, some are fast
- Learn how to recognise a chemical reaction
- Learn the difference between chemical reactions and physical changes

LOOK AT these pictures and decide which substances are made partly or mainly of chemicals.

(The answers are at the bottom of page 15 – but don't cheat.)

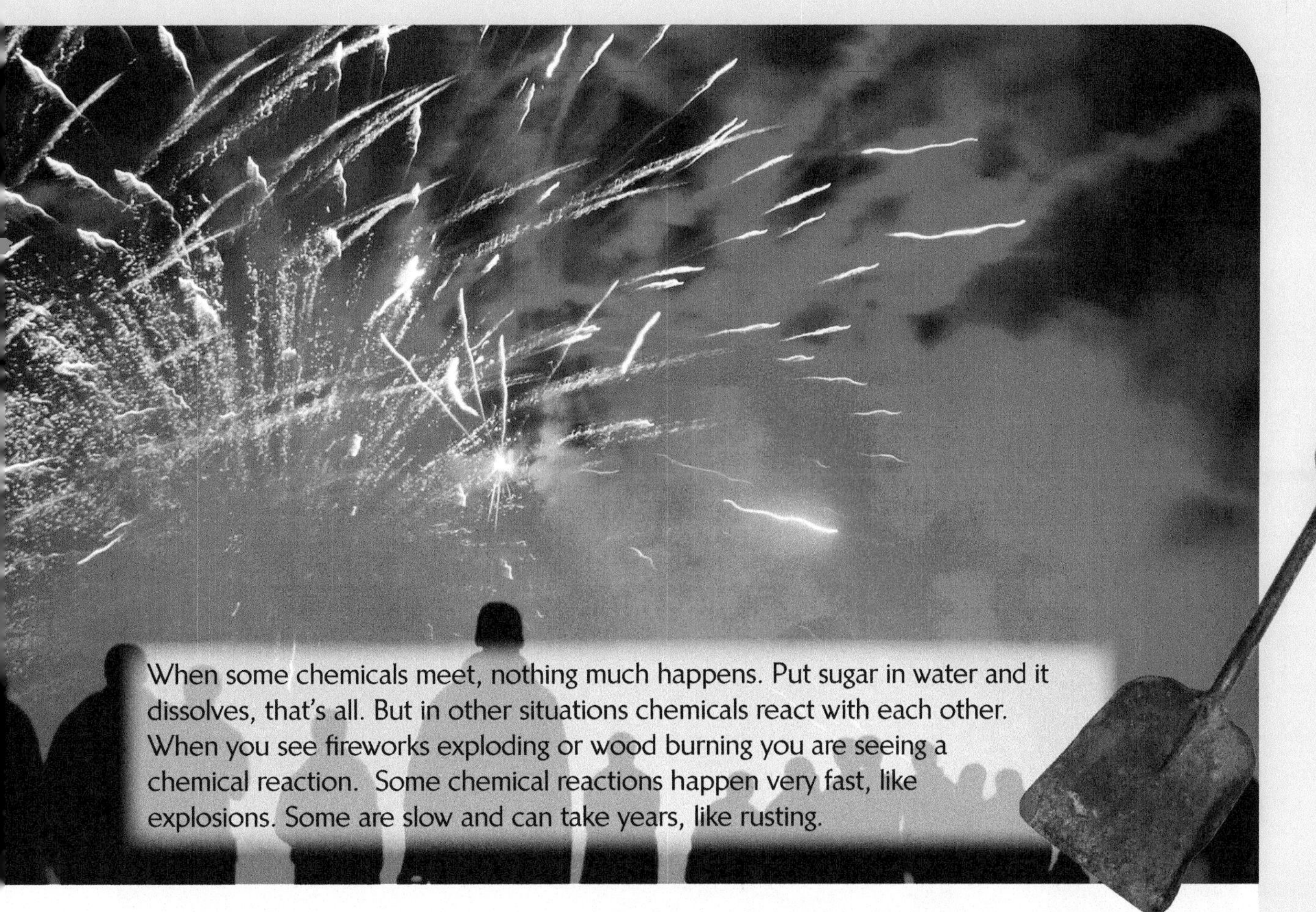

When some chemicals meet, nothing much happens. Put sugar in water and it dissolves, that's all. But in other situations chemicals react with each other. When you see fireworks exploding or wood burning you are seeing a chemical reaction. Some chemical reactions happen very fast, like explosions. Some are slow and can take years, like rusting.

What happens when two chemicals decide to react?

Very often they simply join up with a new partner to form new combinations. For example, let's look at two atomic partnerships.

A and **B** are a chemical atom team		**C** and **D** are another chemical team
	Let's introduce them and see what happens.	
At the start **A B** meets **C D**	→	at the end **A C** and **B D** are the two new teams
  **A B** meets **C D**		 **A C** and **B D**

Important! Did you notice?

No new atoms have been made. No atoms have been destroyed.
These two facts are true about ALL chemical reactions.

The trouble is, you can't see atoms. So how can you tell if two chemicals are reacting or not?

Here are five main clues:

- **Gas.** If there is a lot of gas and bubbles, it is a reaction.
- **Smell.** Some gases are smelly.
- **Colour.** A colour change is usually a clue.
- **Heat.** If things get hot, it's definitely a reaction.
- **Flame.** Every flame has chemical reactions going on.

Fires are very hot chemical reactions, giving off heat and light, with gases reacting together.

Sometimes you get everything at once! A firework reaction produces colour, flame, heat, and an explosion of hot gas.

Physical and chemical

Imagine these two simple actions. First, melt a block of ice. This is a physical change. Second, burn a piece of paper. This is a chemical reaction. This shows us two main differences:

- physical changes **are** easily reversed
- chemical reactions **are not** easily reversed.

Sometimes the new chemical teams made in a chemical reaction are so scattered they seem to have vanished. You can burn a whole pile of wood and end up with just a small amount of ash. Where has the rest of the burned wood gone to? Yes, up into the air. It still exists, but the resulting gas has simply floated away out of sight.

Here is a simple chemical reaction. (Best do it in a sink.) Put some vinegar (a kind of acid) in a test tube or any other container. Get some baking powder and drop it into the vinegar, and see what happens.

Now wash your hands and clean up the mess!

QUESTIONS

1. Write down five clues that tell you that a chemical reaction is happening.
2. Name one difference between a chemical reaction and a physical change.
3. Write each of these in two groups. (One group is headed **chemical reaction**, the other **physical change**.) Ice melting. Egg being fried. Metal rusting. Petrol evaporating. Candle burning. Paint being mixed.
4. Take any three of your answers to the question above, and for each of them, explain the reasons you put it in that group.
5. Name two examples of chemical reactions where light and heat are made.
6. Name one example where a chemical reaction produces a gas.
7. List two things that don't happen to atoms in a chemical reaction.
8. Draw colour-coded atom pictures that show what happens in this chemical reaction: EFG meets JK. They end up making three substances: EK and FJ and E.

Answer: They are all chemical.

Extension:

In a burning candle, what do you think is the main fuel — the wax or the wick? Give a reason for your answer. Plan a hands-on test to check your answer.

with atoms?

In this chapter we will:

- Learn that atoms are one kind of particle
- Learn that particles are always moving
- Learn about the three states of matter: solid, liquid, gas

EVERY OBJECT, substance and living thing is made up of particles too small to be seen. Atoms are one kind of particle. Whether we use the word 'particle' or 'atom' it is a challenge to think about things you can't see.

Sometimes atoms are on their own, but often they join up in groups or teams. For example, water has three atoms tightly joined together: two hydrogen and one oxygen. A group of joined atoms is also known as a 'molecule'. In this chapter we will use the word 'particle' whether they are atoms or groups of atoms. Whatever their name, particles are small. A single drop of water has more particles in it than the human population of the entire world!

one oxygen atom

two oxygen atoms make one molecule

one molecule of water – one oxygen and two hydrogen

one molecule of propane, or bottled gas (the three bigger atoms are carbon)

Solid, liquid and gas are called the three 'states of matter'.

Take water for example.

- Water can be a solid – we call it ice.
- Water can be a liquid – we call it water, or wai.
- Water can be a gas – we call it steam or water vapour.

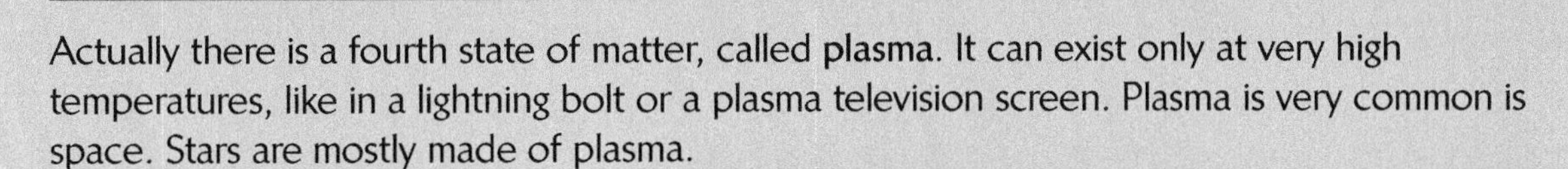

Actually there is a fourth state of matter, called **plasma**. It can exist only at very high temperatures, like in a lightning bolt or a plasma television screen. Plasma is very common is space. Stars are mostly made of plasma.

You can think of particles of the three states of matter as being a bit like humans behaving in different ways. Particles are busy little things, constantly on the move.

- Particles move slowly in a solid, like ice.
- Particles move faster in a liquid, like water.
- Particles move even faster in a gas, and are spaced wide apart.

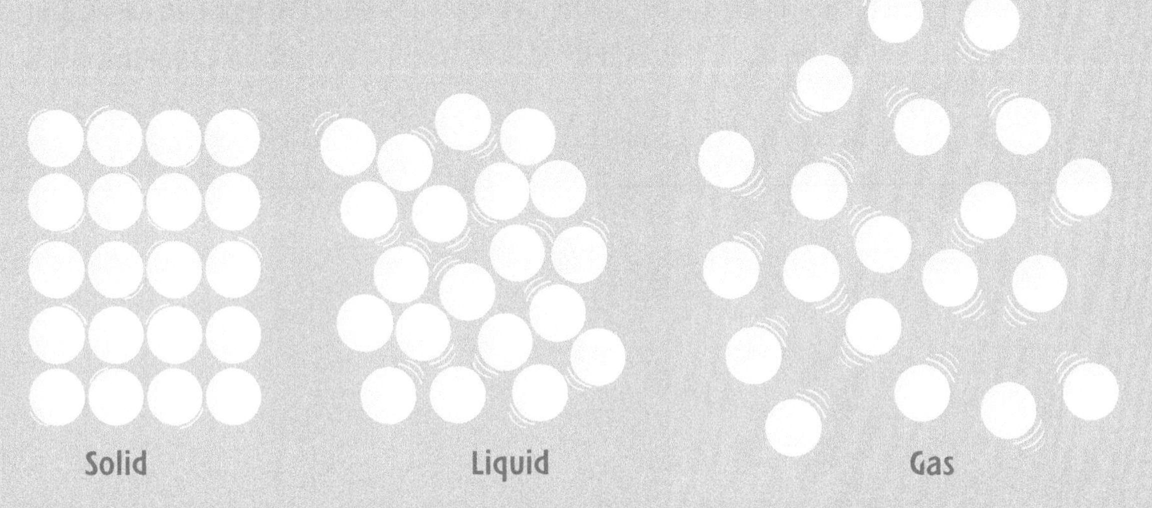

The particle idea helps to explain many things about solids, liquids, and any gas.

Solids

Any solid has:

- a definite weight
- a definite shape
- a definite volume.

Why are solids a definite shape? Because the particles are all firmly linked to their neighbours, and are not free to move around very much.

Liquids

Any liquid has:

- a definite weight
- a definite volume
- no definite shape.

Why does a liquid take on the shape of whatever it is poured into? This is because the particles are not firmly tied to their neighbours, and are free to move around.

Gases

Any gas:

- has a definite weight
- does not have a definite volume
- does not have a definite shape.

Why are gases so 'light'? This is because their particles are moving fast with wide spaces in between.

You can't see a gas, but you sure can feel it when it pushes against you.

Hands-on

Start with a litre of water.

Pour it into four different containers one after another without spilling a drop, then back into the original bottle.

What did you notice about the change in shape?

Did the volume increase or did it stay the same?

For both these questions, write down what you saw.

QUESTIONS

1. Say if each of the following five statements is true or false:
   - A desk is made out of atoms.
   - Gas is a common state.
   - Water can be found in all three states.
   - Atoms are constantly moving.
   - Stars are made up of plasma.
2. Complete the following tasks. (Copy the table.)

State	How fast do the particles move?	How are the particles arranged?	Does it have a definite volume?	Does it have a definite shape?
Solid				
Liquid				
Gas				

3. What is the fourth state of matter known as? Where does it occur on earth? Where does it occur in space?
4. How many atoms in a molecule of oxygen?
5. How many atoms in a molecule of water?
6. How many atoms in a molecule of propane?

Extension:

Use the library or Internet to find out the molecule structure of water, CO_2 and octane. Use polystyrene balls to make models of each. Paint the 'atoms' before you glue them together.

ELECTRIC effect

In this chapter we will:

- Learn that electricity is a form of energy
- Learn that electrical energy can be changed to other kinds of energy
- Learn about different kinds of circuits
- Learn how electricity is made
- Learn what kinds of substances are good conductors

LIFE WOULD be very different without electricity, right? Let's look at the main differences between static electricity and circuits, and how electricity is made.

There are different kinds of energy such as heat, light, sound, movement, and electrical energy - also known as 'electricity'. Sometimes electricity is also called 'power', although that is not quite correct.

LOUDSPEAKER

electrical energy changes to sound energy

HEATER

electrical energy changes to heat energy

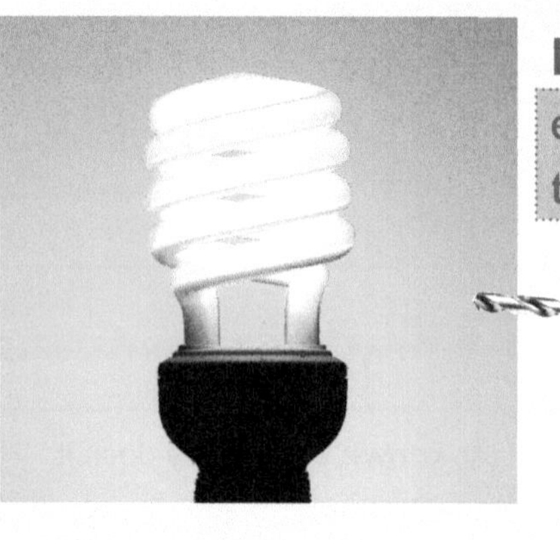

LIGHT BULB

electrical energy changes to light energy

POWER DRILL

electrical energy changes to movement energy

There are two types of electricity:

- static (which mostly stays in one place)
- current (electricity moving fast).

Static electricity

Combing your hair with a plastic comb can make static electricity. So can stroking a cat. What happens is that some electrons (negative particles inside atoms) get rubbed off your hair and give the comb a slight negative charge. A lightning strike is caused by a huge build-up of static electricity in a cloud.

Current electricity

When you switch on a lamp, or television, or a torch, you are using current electricity.

Let's have a closer look at a torch.

metal spring
C size cells connected in series
lamp contact
lamp
reflector
slide switch
metal switch contacts
plastic casing

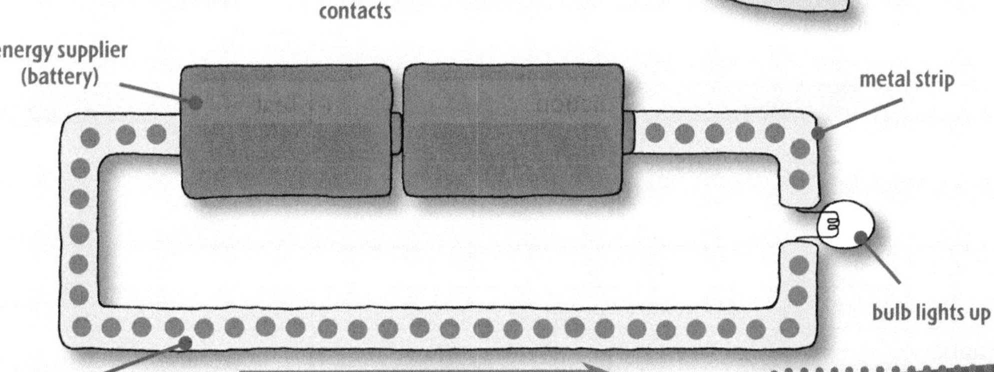

I'm feeling negative, so I must be charged-up

What happens?

Electrical current flows from the negative end of the battery in a circuit (circle) around to the positively charged end. The circuit has an energy supplier (the battery), and an energy user (bulb), and the conductors (wires).

Circuits

A circuit is an arrangement of wires that keeps an electric current flowing. You can't see the current, but you can imagine it like water going through pipes. The battery gives the 'push'.

There are two main kinds of circuit: series and parallel circuits.

Series circuits

Series circuits have no branches. The current can only follow one path. If you turn the switch off, both bulbs stop working. If you break one bulb, the other one goes out.

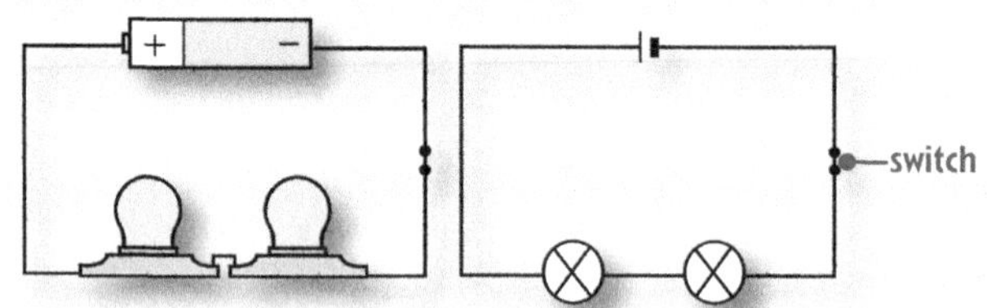

Parallel circuits

A parallel circuit has more than one path, with the different paths running side by side. If one bulb breaks or is switched off, the other bulbs will still glow. Switches can be put into a circuit to switch on all parts, at once, or just one or two parts.

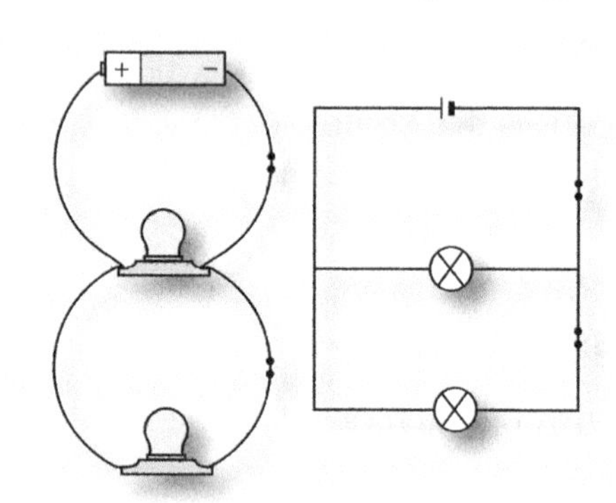

Hands-on

Your teacher will give you information on how to test different substances to see if they can conduct electricity.

Set up a series circuit like this diagram shows. Join the two bare wire ends A and B. Make sure the bulb lights up. Next, try putting different materials in the A-B gap, one at a time. If the bulb lights up even a little bit, that shows the material can carry electricity: it is a conductor. If the bulb does not light up, the material is a non-conductor, also known as an insulator.

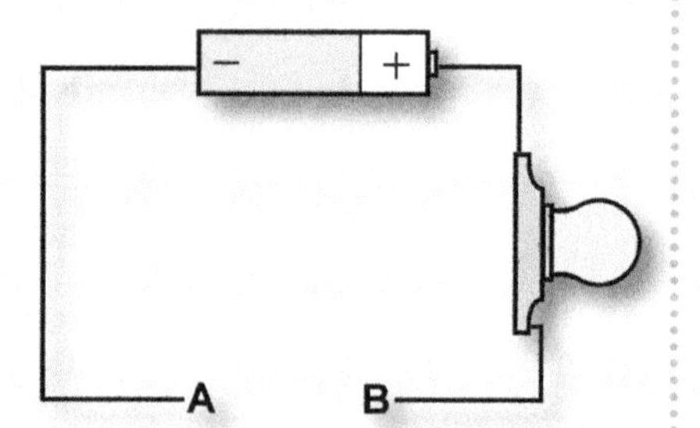

Material	Prediction	Test
Scissors		
Pencil graphite		
Plastic		
Coin		
Wood		
Fabric		

What are the main ways of making electricity?

Power stations make electrical energy. They start off with some other kind of energy. It can be a chemical (usually fossil fuel: coal, or oil, or natural gas), or wind, or water pressure, or sunlight. About 70% of New Zealand's electricity is from 'hydro power', made by water flowing through pipes to drive turbines.

Thermal power station

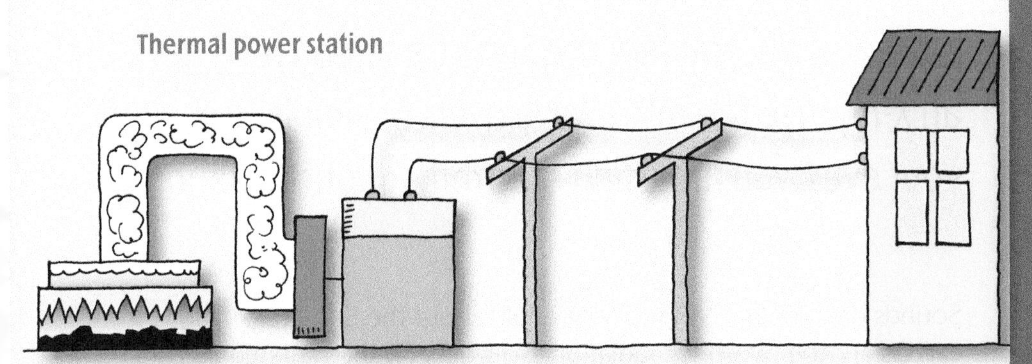

In a thermal power station (above), the heat from burning coal or gas is used to boil water. The steam from boiling water is used to drive turbines and generators. Huntly power station has four generators each making 250 million watts of electric power.

A hydro power station.

Solar-electric panels turn sunlight into electric energy.

Hands-on

Make a circuit. It must have at least one switch and one bulb in it. Make a simple drawing of your circuit. You could design your own circuit or make one of the circuits drawn in this chapter. If your circuit has two batteries (cells) find out if it makes any difference which way round they are connected.

QUESTIONS

1. Make a list of at least ten different things in your house which use electrical energy.
2. Draw a series circuit which uses one cell and two light bulbs. Show clearly how wires join on to the other parts.
3. Now, draw a parallel circuit which uses one cell and two light bulbs. Show clearly how wires join on to the other parts.
4. Add to your parallel circuit drawing one switch which will turn both bulbs on at once. Also add a second switch which will turn on just one of the bulbs.
5. Do you think all the mains electricity appliances in your house are arranged in a series or parallel? Give a reason for your answer.

Extension:

1. Find out the parts of an electric motor and how the motor works.
2. Find out how most of the country's electricity is made, and where.

MICROWAVES, UV, TV

In this chapter we will:

- Learn what radiation is?
- Learn about different kinds of electromagnetic radiation
- Learn about the uses and dangers of electromagnetic radiation

EVERY TIME you feel the Sun's heat, watch TV, get an X-ray, use the microwave oven, or simply use your eyes, you are using electromagnetic radiation.

Sounds dangerous? Not really. We talk about the Sun's rays. 'Ray' can be another word for radiation. So we mean the same thing whether we talk about electromagnetic waves, rays, or radiation.

TV and microwaves and X-rays and UV and visible light all share one important feature: **they all move at the speed of light.**

The electromagnetic spectrum

There are many kinds of radiation that move at the speed of light. They form a spectrum (that means a whole range) from short wave-length to long wave-length. The only parts we can detect with our own senses is visible light (with our eyes), and infra-red heat (with our skin).

Light is seriously fast! It takes only 8 minutes to travel 150 million kilometres from the Sun to me!

SHORTEST WAVELENGTH → LONGEST WAVELENGTH

X-rays	UV light	Visible light	Infra-red	Microwaves	Shorter wavelength radio	Longer wavelength radio and TV

Plants use light energy to grow. We can also capture some of it with solar panels. The radiation from the Sun is a mixture of harmful rays (like UV), visible light and infra-red.

Let's check the ways in which all radiation types are the same.

- They all move at the speed of light, 300 000 km a second.
- They can move through empty space.
- They all move in straight lines.
- They can all be reflected (bounced) and refracted (bent).
- They have energy but no substance.

Radio waves

Radio waves are what mobile phones and radio and television stations send out. Stars also send out radio waves. Radio wavelengths are the longest of all: from five centimetres to over a hundred metres long. We can't hear them. Phones, radios or TV sets first have to pick them up and convert these electromagnetic waves into sound waves which we can hear. Sound waves are completely different from electromagnetic waves.

Microwaves

Microwaves can be used to cook food, but only if the food contains some water. Planes and ships use similar wavelengths for radar. Microwaves are 1 cm to 5 cm long.

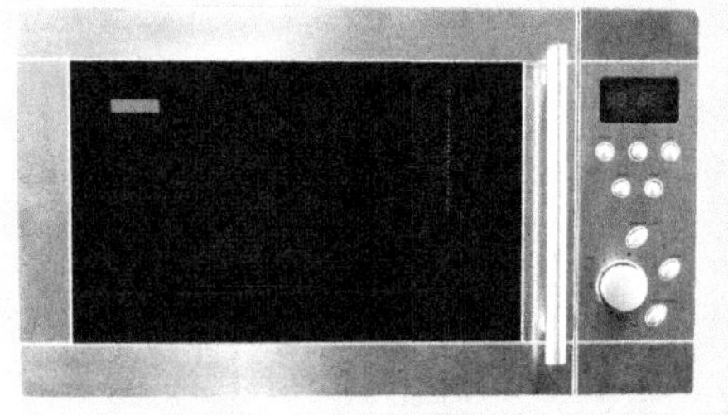

Infra-red light

We can't see this, but we can feel it on our skin as warmth.

Visible light

This is the part that we can see, and also the part that green plants use for energy. The visible spectrum has wave lengths from red to blue. Visible wavelengths are less than 1/1000 mm long.

Ultraviolet (UV)

The Sun gives off plenty of ultraviolet radiation (UV), which in big doses is dangerous to living things. Fortunately, the air has a thin layer of ozone, which screens out most of the UV. If it wasn't for this protection, life would not survive. Many insects can see UV light.

X-rays and gamma rays

These have very short waves with a high frequency, and are highly dangerous in big doses. That's why your dentist goes out of the room when you have an X-ray! Gamma rays are given off by the sun, and also by some kinds of radioactive elements like plutonium.

Use a prism to show that white light is actually a mixture of different colours. (Your teacher will show you how.)

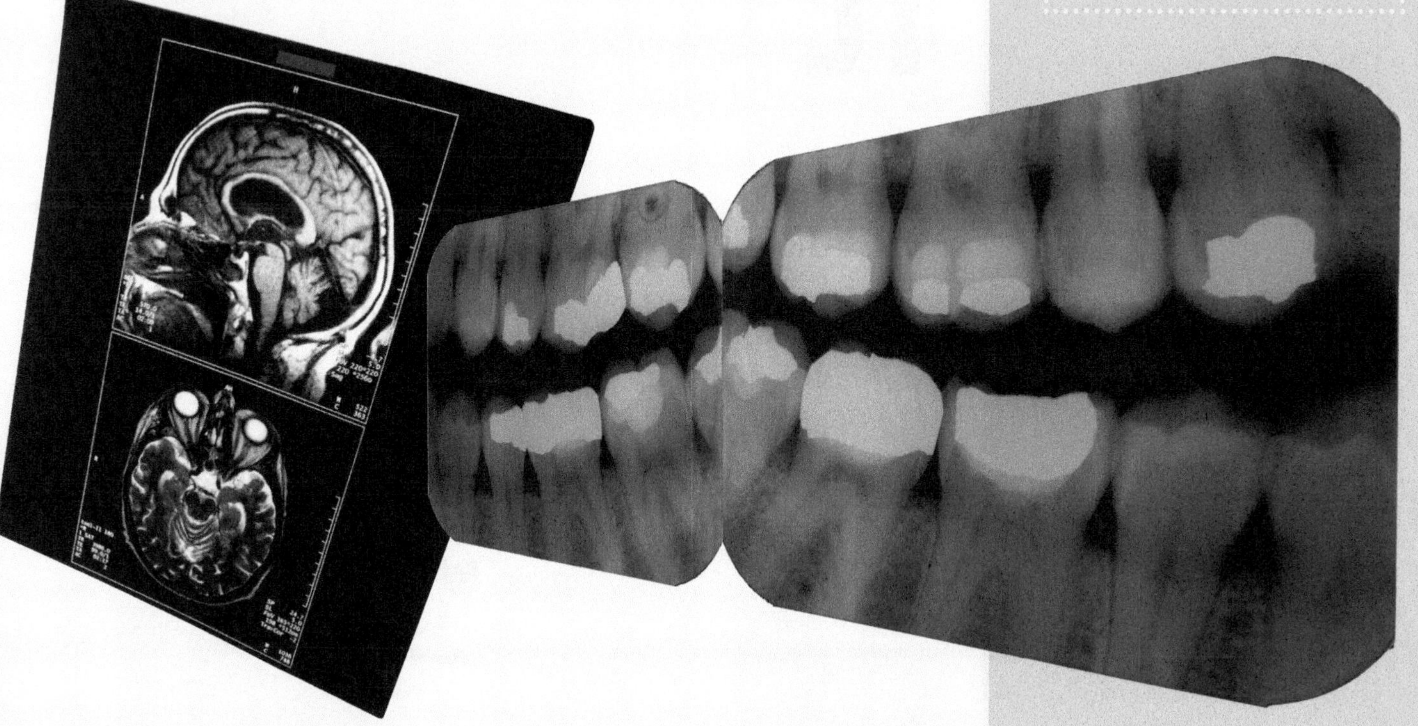

QUESTIONS

1. What name do scientists give the family of waves that move at the speed of light?
2. What kinds of electromagnetic radiation have the shortest wavelength?
3. What kinds of electromagnetic radiation have the longest wavelength?
4. Name four types of radiation that are given off by the Sun.
5. Copy and complete this table.

Type of electromagnetic wave	Does it get through the Earth's atmosphere?	How dangerous is it to living things?	Practical uses in everyday life
X-rays			
Ultraviolet			
Visible light			
Infra-red			
Microwaves			
Radio waves			

Extension:

Investigate a famous nuclear accident like Chernobyl, and the impact on the environment and people.

FORCING it

In this chapter we will:

- Learn about different kinds of force
- Learn that unbalanced forces cause movement
- Learn that forces work in pairs
- Learn about simple machines like levers and ramps

A FORCE is a push or a pull. The wind pushes against you, a dog pulls on its lead, a magnet pushes away another magnet.

As far as we know, there are only four kinds of physical force in the universe:

- contact
- gravity
- magnetism
- electrostatic.

Electrostatic

Gravity

A simple action like riding a bike involves many different contact forces, like when you speed up, turn corners, stop or go up hills.

When you go uphill, the pull of gravity means that you have to push harder on the pedals to keep moving. If your push is too weak, the bike will slow down and stop. When you go downhill, the force of gravity helps you.

Have a careful look at the drawing below. Joe's hand is pulling on the rope – and the rope is pulling on Joe's hand. The two forces are shown exactly equal, and the boat is not moving.

What will happen if Joe pulls harder?

That's correct: the boat will move up the slope, perhaps even speed up. What will happen if Joe lets go? Correct again: the boat will move down the slope and pick up speed.

The boat example on page 29 shows us two things:

- forces act in pairs, and each force has an opposite force
- unbalanced forces cause movement.

You can picture the same sort of situation with dogs and their handlers.

In many cases, it is impossible to lift a heavy object straight up. A ramp – also called an 'inclined plane' – can make the job easier. A lever is another way of moving heavy objects.

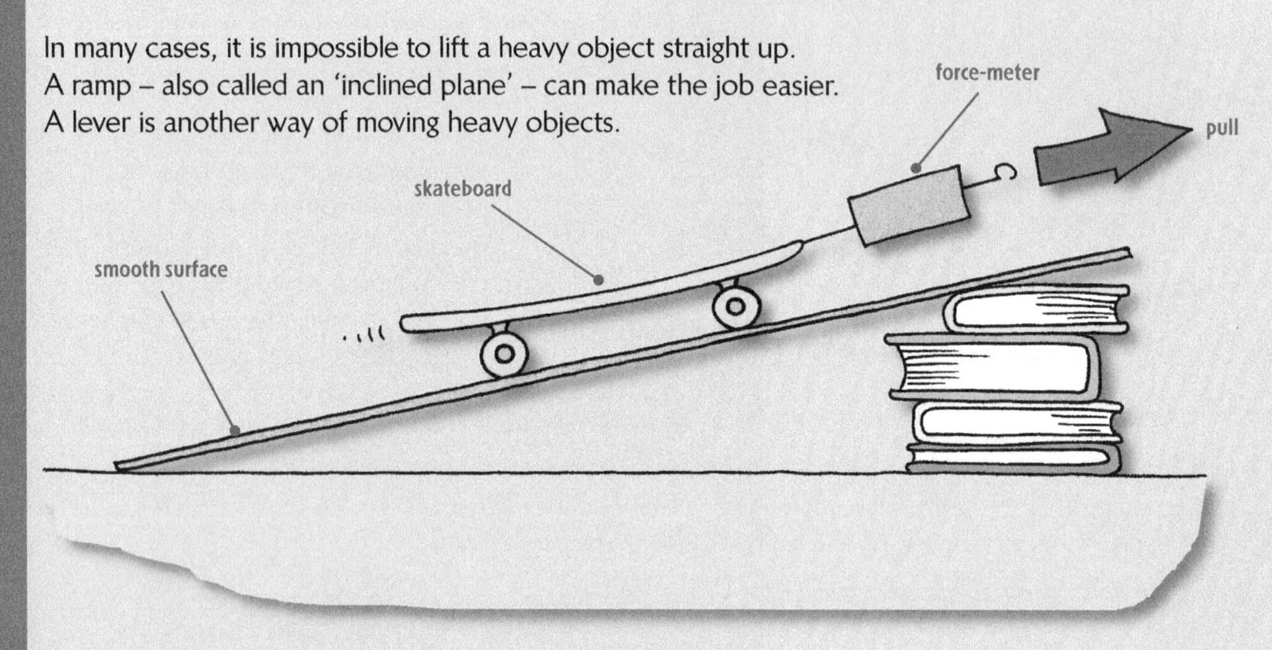

Anything that helps a small force overcome a big force can be called a 'machine'. Inclined planes and levers are simple machines. A pulley is another example of a simple machine.

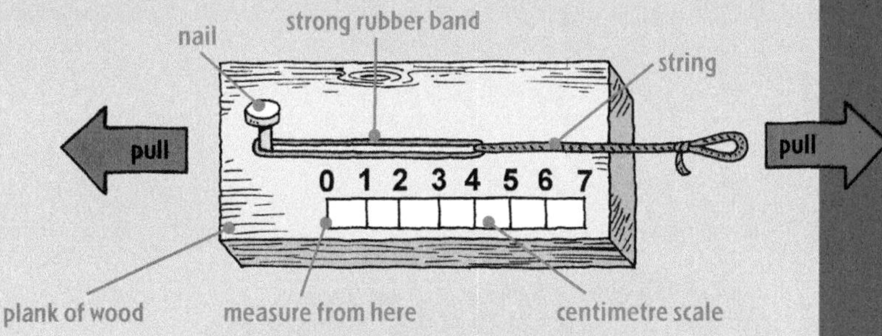

Aim: to measure the friction of different surfaces.

- Make your own force-meter like the drawing shows, or else you can use a bought one that measures in grams or kg.
- Use your force-meter to drag a bag across the floor.
- Try dragging it across different surfaces, such as carpet, wood, concrete.
- In each case write down the average pull needed to keep the bag moving slowly.
- You could also use the force-meter to measure the force needed in different situations, such as lifting objects, or turning a door handle.

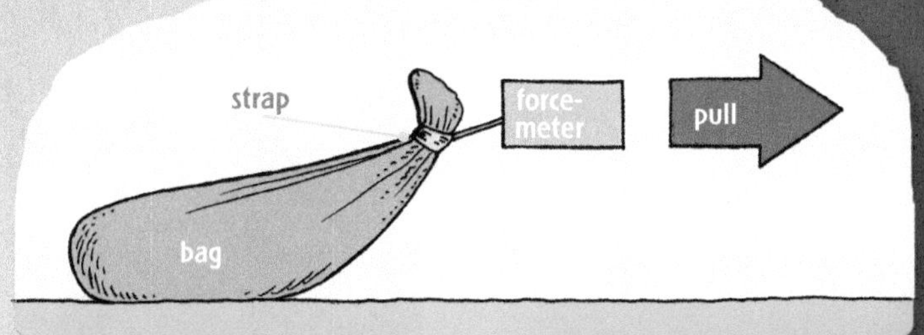

QUESTIONS

1. List the four kinds of force.
2. Copy and complete these three sentences:
   - A force is a ..... or a .....
   - Forces always exist in .....
   - Unbalanced forces cause ......
3. When the moon pulls on the tide, what kind of force is this?
4. In each of these drawings the arrows show the force of a dog pulling on its lead, and the owner's hand pulling in the opposite direction. In each case say what movement will happen to the left or right.

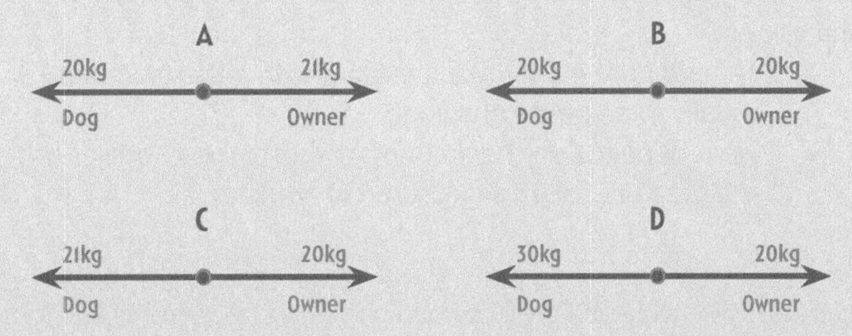

5. Name three kinds of simple machine named in this chapter.

Extension:

A 10 kg pull can lift a 20 kg weight, a 30 kg weight, and even a 100 kg weight. Find out how 20 kg, 30 kg and 100 kg can be lifted using only pulleys and a strong rope. Make drawings of these three situations.

Hint: an experienced sailor may be able to give you advice on this.

WHAT'S hot?

In this chapter we will:

- Learn that heat and temperature are not the same
- Learn what makes things hot
- Learn how heat gets from one place to another

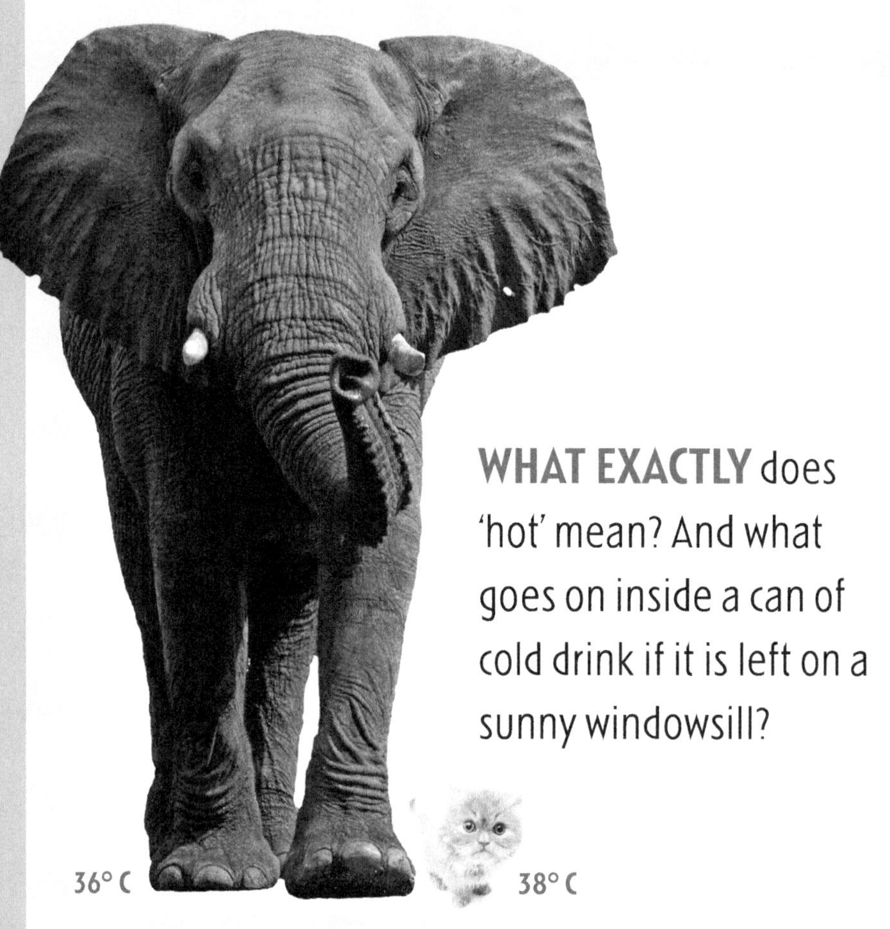

WHAT EXACTLY does 'hot' mean? And what goes on inside a can of cold drink if it is left on a sunny windowsill?

Look at this situation: an elephant at 36° C body temperature, and a cat at 38° C body temperature.
Who has the highest temperature?
That's right, the cat.
Whose body contains the most heat (warmth)?
That's right, the elephant.

You have just proved to yourself that heat and temperature are not the same.

- Temperature is what you measure with a thermometer, usually in degrees centigrade.
- Heat depends on the **amount** of substance as well as its temperature. Heat is measured in 'joules'.

Leave a bucket of hot water to stand for an hour, and what will happen? Right again. It will cool down, losing heat and falling to a lower temperature.

So what actually goes on inside something when it is hot? It's those billions of invisible particles. When something is at a high temperature, the particles move faster, like children at playtime, and there is lots of energy. At a lower temperature, everything is slower and there is less energy.

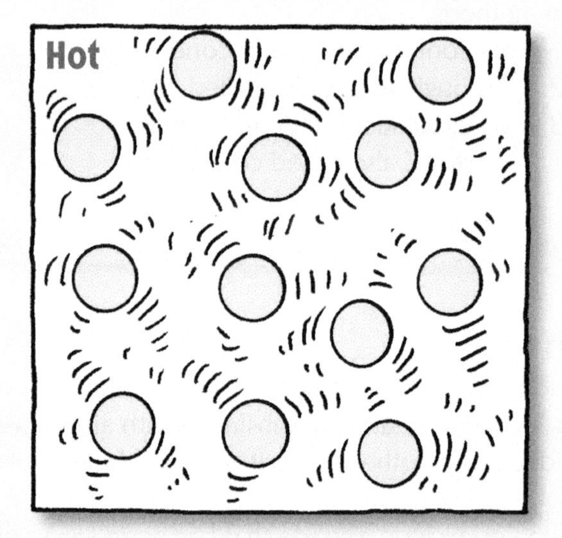

All sorts of things give off heat: a stove, radiator, electric light, computer, your own body. But none of these actually make energy. In every case they have got the energy from somewhere else. The biggest energy source of all is the Sun. If the Sun was switched off, the sea and every living thing would soon be solid ice. But don't worry, the Sun has been going for about five billion years and will keep going for billions more.

Did you know?

When you are sitting or resting, you give off about the same amount of heat as a light bulb: about 60 joules each second.
When you are running, your heat output goes up to 300 joules a second.
Your normal body temperature setting is 37° C.
If your body temperature falls to 32° C you will become unconscious.
If your body temperature rises to 42° C your life is in danger.
Water boils at 100° C, but at high altitude it boils at 90° C or less.

No danger of me overheating!

Warm things, for example a penguin or a warm building, can't help giving off heat to their colder surroundings. How does heat get from place to place? There are three main methods.

1. **By contact**, also called **conduction**.
2. **Through space**, also called **radiation**.
3. **By circulation** of air and water currents, also called **convection**.

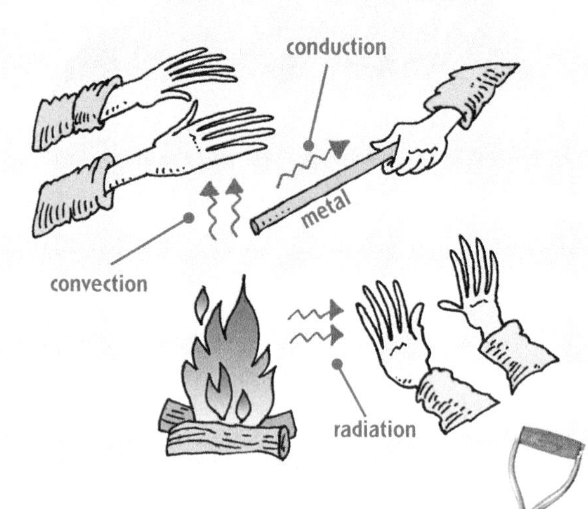

Conduction

Conduction of heat can happen only if there is contact. Metals are the best conductors of heat. Any substance with air spaces inside is a poor heat conductor – in other words it is a good heat insulator.

When you pick up a spade that has been lying out overnight, all parts of the spade are at the same temperature but the metal part feels cold to touch, and the wooden handle feels less cold. This is because heat flows from your hand into the metal (a good heat conductor), but flows slowly into the wood (a poor conductor).

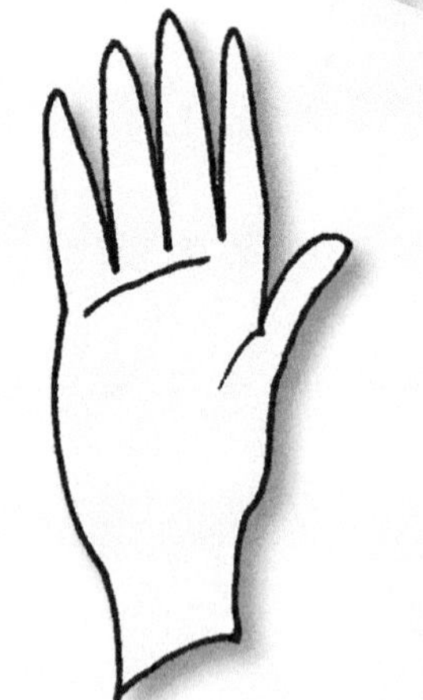

Radiation

Hot objects such as irons and electric bulbs give off infra-red radiation in all directions. You can't see this infra-red heat, but you can feel it from a distance.

Hands-on

Take a long piece of wire and drip some candle wax half way along it. Then wait until the wax is cold and solid again. Using some kind of wooden handle at your end, put the other end in a flame. Wait to see what happens.

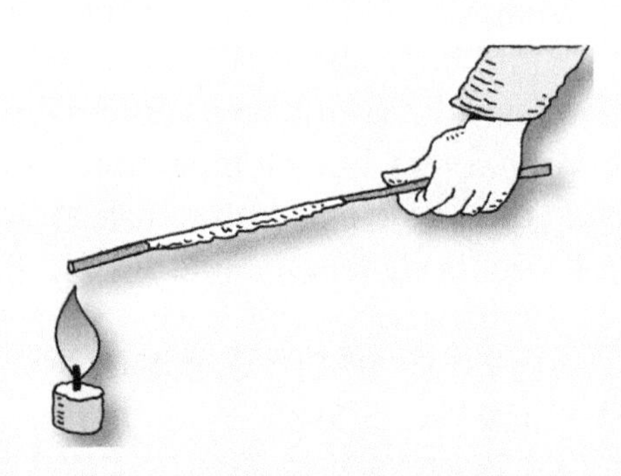

Convection

Hot air rises and cold air falls. Because hot air is less dense (lighter) than cold air, the air around the heater slowly rises and cold air moves in to take its place. This sets up a circular movement of air called a convection current. The same sort of currents happen in water being heated on a stove.

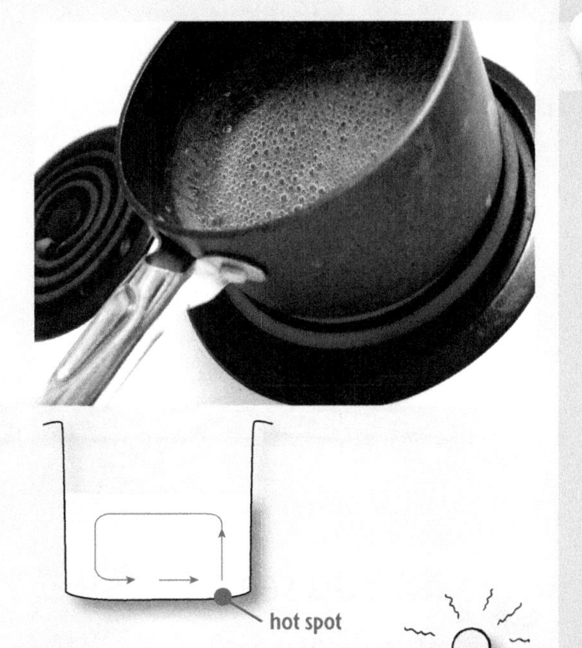

Colour matters ... sometimes

The colour of a surface makes a difference to how fast it loses or gains heat by radiation. Light and shiny objects lose heat slowly and also gain heat slowly. Dark objects lose heat fast and also absorb heat fast. This only applies to radiation. Conduction and convection are not affected by colour.

Hands-on

Buy a helium party balloon, then add small bits of plasticine to the string until the balloon is 'weightless' and balanced and will stay hanging in the air wherever you put it. (It helps if all the doors and windows in the room are shut!) Place the balloon above something warm — like a heater — and watch what happens next. Draw the balloon's travel around the room.

QUESTIONS

1. What is your normal body temperature? What is your heat output when you are resting? And when running?
2. Describe what happens to the metal atoms in the hot plate on a stove, as it warms up from room temperature to 200 °C.
3. Which will heat a hangi better: ten stones at 200°C, or 20 stones of the same size at 200°C? Explain why.
4. In your own words explain the difference between heat and temperature.
5. Name the three ways heat escapes from a kettle of water when it cools down from 100°C to room temperature.
6. Explain why kettles are mostly white or shiny in colour.
7. On a sunny day, a black T-shirt feels hotter to wear than a white T-shirt. Explain why.
8. On a cold cloudy day, a black T-shirt feels colder to wear than a white T-shirt. Explain why.
9. Gloves keep your hands warm. Name three ways they help do this.
10. 'Pink Batts' help trap heat. Explain how they work.

Extension:

What temperature is absolute zero? Why is it called absolute? Strange things happen at absolute zero — what are they? And what temperature does water freeze and boil at on the Kelvin scale?

SEEING the light

In this chapter we will:

- Learn that light can be reflected and refracted
- Learn that white light is a mixture of colours
- Learn about convex and concave lenses

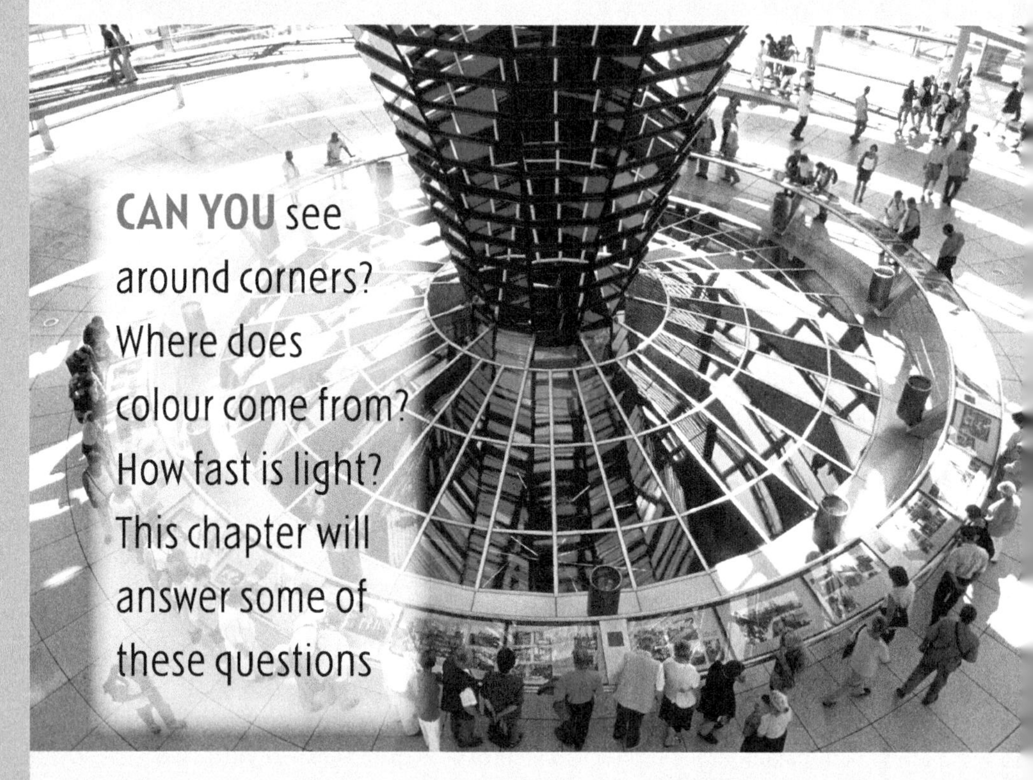

Light bounces

When light hits a surface some of it is absorbed, but some bounces off. Another word for 'bouncing off' is 'reflection'. If the surface is rough, the reflection is scattered. If the surface is shiny and smooth – like glass, polished metal, water – the reflection will be 'perfect' like in a mirror. If you look carefully at this kind of reflection you will notice some interesting things:

- the incoming angle is always equal to the outgoing angle
- the image is turned around.

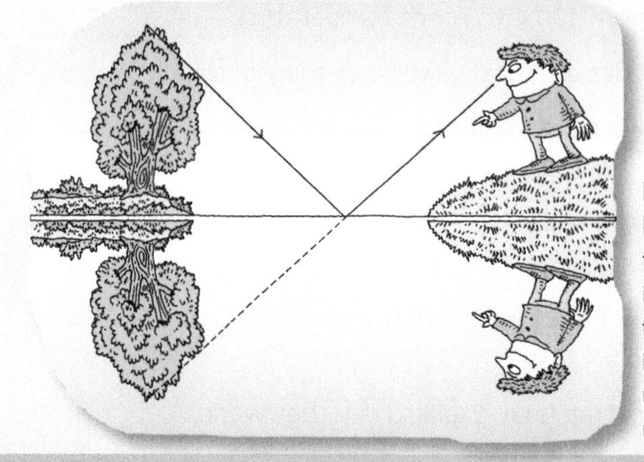

Because your mind knows that light travels in straight lines it **looks** as though the reflected light is coming from an upside-down tree under water.

Light bends

When white light goes from one transparent substance (see-through, like water, air and glass) to another, it changes direction. This bending is called **refraction**.

The angle is important.

Light refracts when it enters glass or water at an angle.

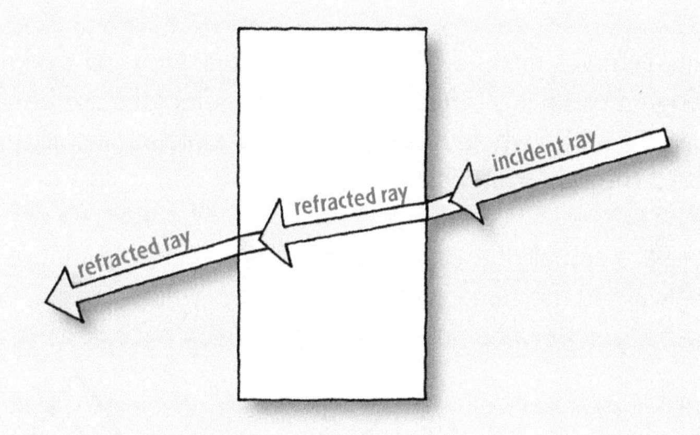

If the light hits the surface at 90°, it goes straight in without refracting.

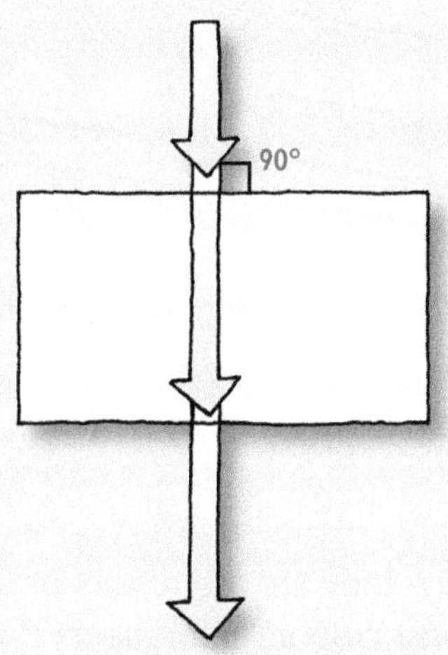

Hands-on

Place a pencil in the centre of a beaker of water and look at it from the side. Now place the pencil to the left or right of centre. Describe what you see, then explain why this happened.

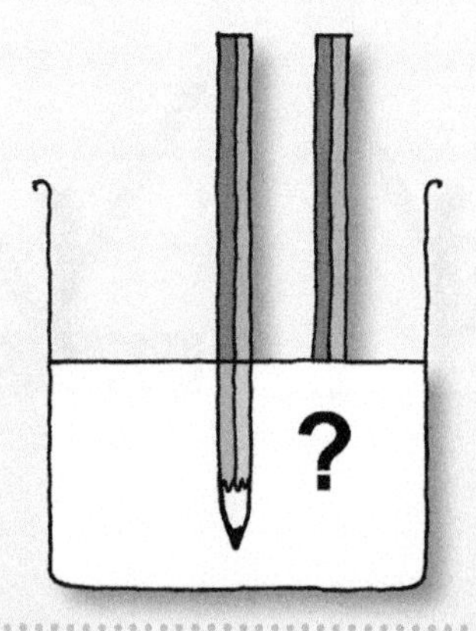

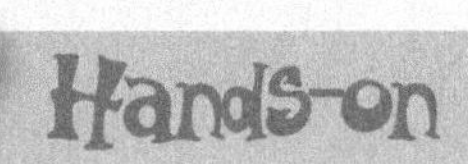

Turning it around. If you look at writing in a mirror you can see that everything is turned around. But if you look at your face in a mirror it looks normal. Explain why. Now place a small hand mirror on this page just above these printed words:

mirror here

CHOICE GRAPES.

What do the words read like in the reflection? See if you can explain why!

How about some CHOICE FISH ?

Colour

White light is a mixture of different colours. How do we know this? When white light travels through a prism it is refracted. Some wavelengths are bent more than others. The result: light separates into a whole range of different colours, just like you see in rainbows. It works just the same if the light is artificial.

There are actually thousands of colours in a spectrum, but we often say there are seven. They are always in the same order: red, orange, yellow, green, blue, indigo and violet.

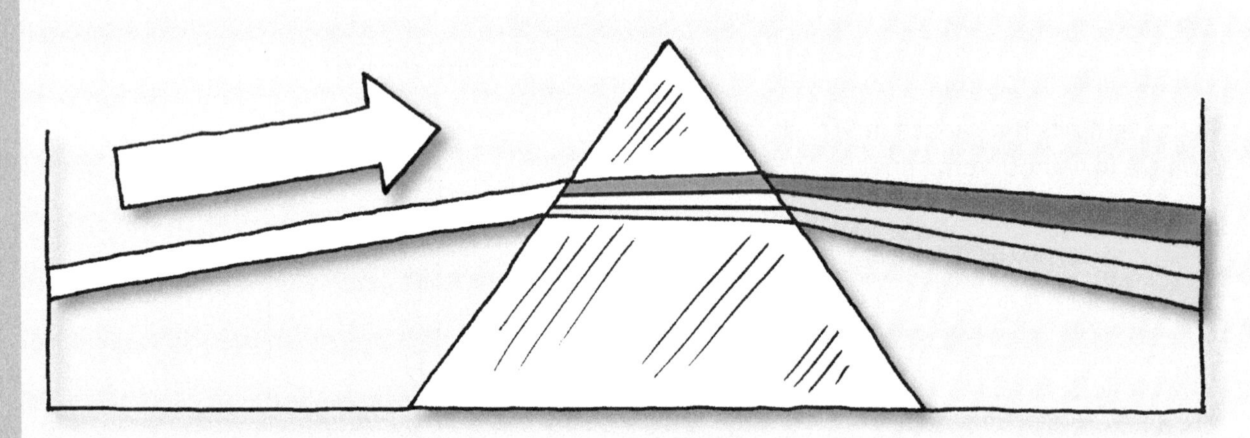

What about ordinary colours such as you see now on this page? Most colours are not made by a prism or a rainbow – they are the results of what happens when light hits a surface. Any coloured surface absorbs most colours and reflects only its own colour. A white surface reflects back whatever light shines on it. A green surface reflects green light but absorbs all the others. A pure black surface absorbs everything and reflects nothing.

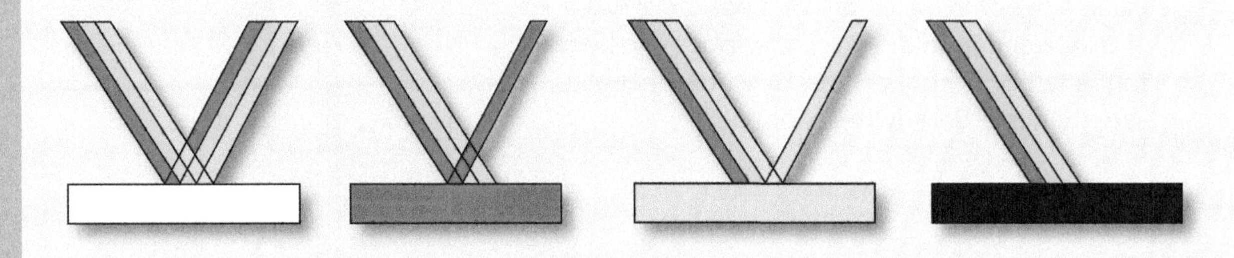

Lenses

A lens has a curved surface that bends light for a special purpose. It can be made of glass, plastic or any other transparent (clear) material.

Convex lens

A convex lens refracts light inwards to a 'focal point'. Seen from the correct distance, a convex lens makes objects look bigger. Each eye has one convex lens inside it.

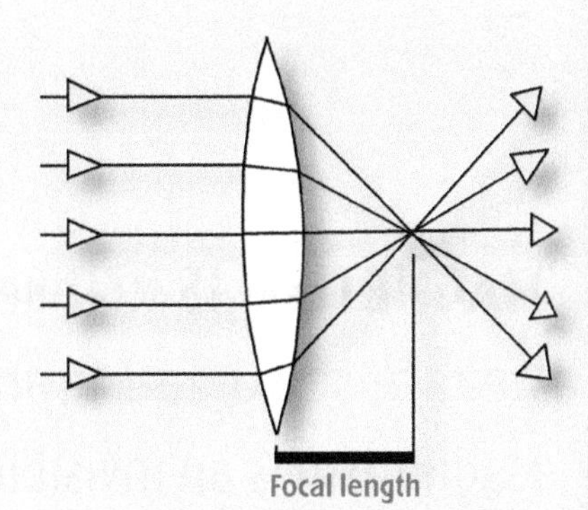

Concave lens

A concave lens refracts light outwards. It will always make objects look smaller. Microscopes and cameras have a combination of convex and concave lenses.

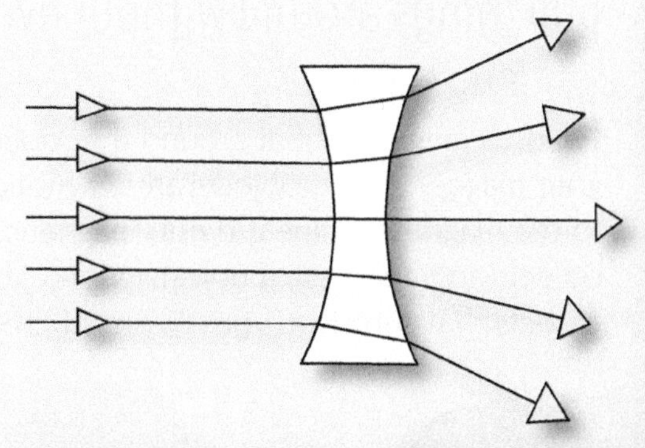

Hands-on

Get a convex lens. Find out its focal length by using sunlight and a ruler and a piece of paper. Next, try holding the lens at a different distance from your eye and looking at different objects. Sometimes the objects look bigger, sometimes the right way up, sometimes upside down. If you can, see how these effects link in with the focal length of your lens. (Warning! Never look at the Sun, either directly or through a lens.)

QUESTIONS

1. Give another word for 'bouncing light'.
2. Give another word for 'bending light'.
3. Draw a concave lens.
4. Draw a prism, splitting white light into colour.
5. Draw a convex lens.
6. List the seven colours of the rainbow.
7. Which of these diagrams is correct? Which ones are wrong? Say why.

8. Which is these diagrams are correct? Which are wrong? Say why.

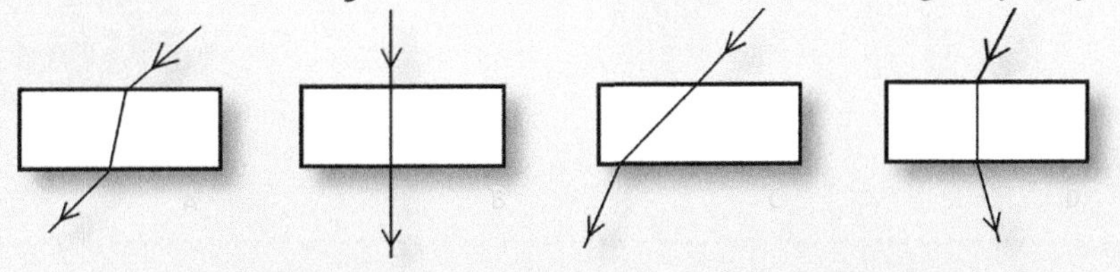

9. Say what colour a red shirt will look like in each of these three situations: in white light, in red light, in blue light.

Extension:

Find out about the structure of the eye in detail, and make a drawing of it. What does the retina do? How does the iris work?

MAGNETIC attraction

MAGNETISM IS strange. It's all around us, but we can't see or feel it, and it works at long range. Magnetism is an invisible force that can push and pull things around without even touching them.

Go on a magnetic hunt around the room, finding what things stick to your magnet. Some objects are magnetic because they are attracted to the magnets. Some are non-magnetic. Make one list of magnetic objects and another of non-magnetic objects. Were there any objects that surprised you? Were all the metal objects magnetic?

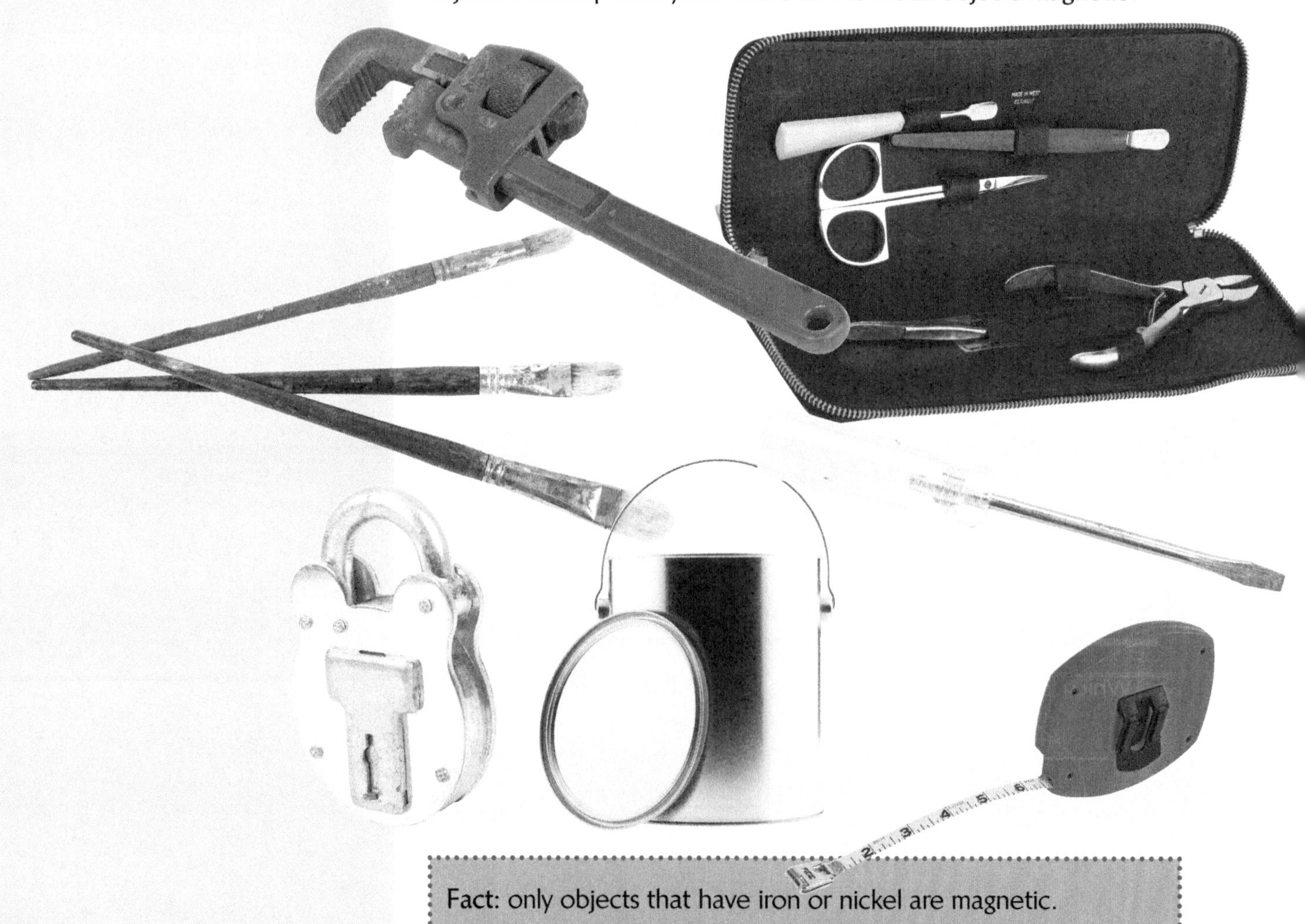

Fact: only objects that have iron or nickel are magnetic.

What happens when you put two magnets together? It depends on what ends you use. These ends can be called 'north' and 'south', but don't worry about this – it is simpler to call them red and blue. Use the word 'attract' for 'pull together', and the word 'repel' for 'push apart'. Question: What two simple rules can you see from your results?

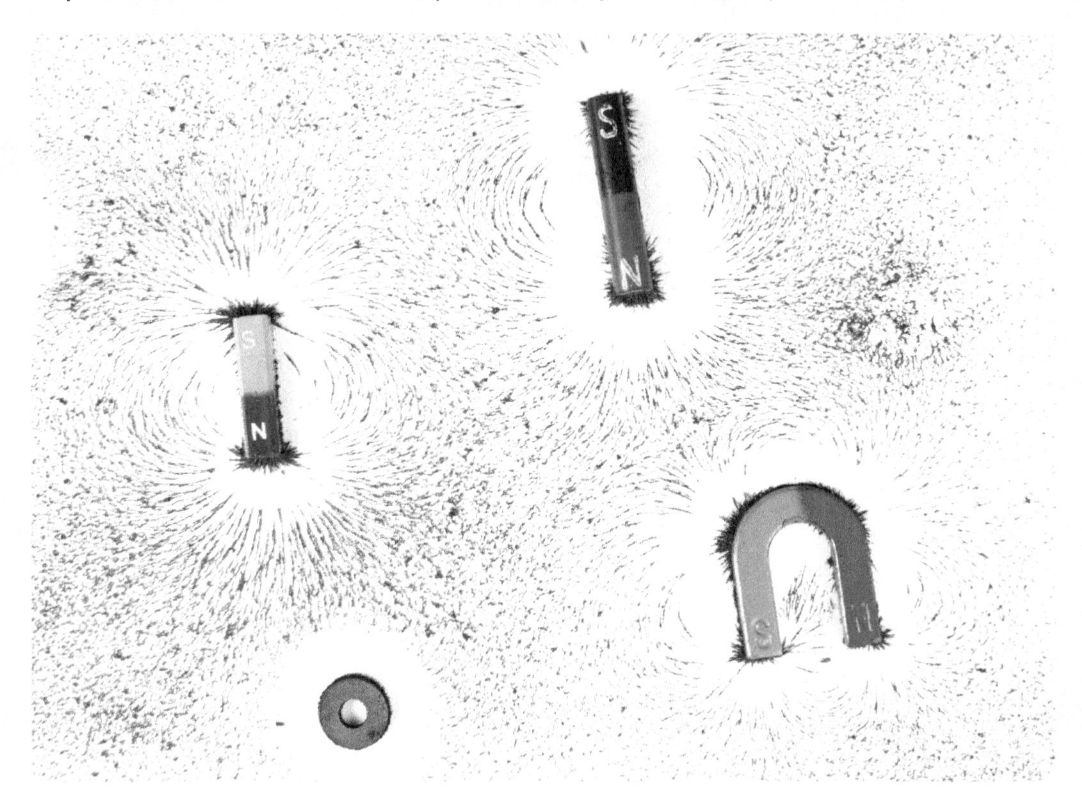

Invisible magnetic fields

Any magnet has an invisible 'force field' around it. If you cover a magnet with smooth flat paper and spread some iron-sand on the paper, then tap the paper lightly, you will see patterns like the ones shown above. Note that it is best to keep the magnets under the paper.

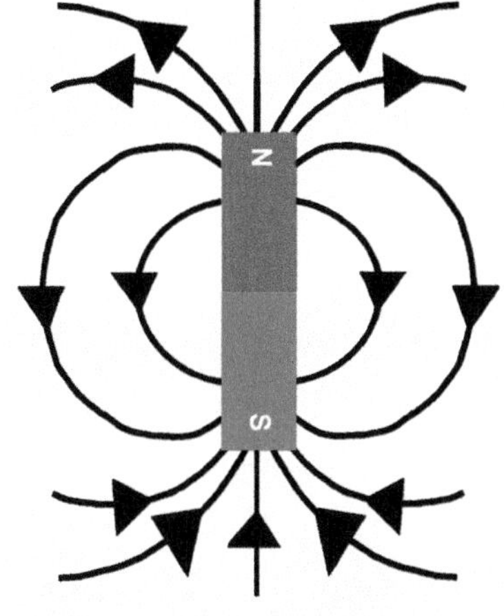

We can detect invisible magnetic fields with iron sand. (The Teacher's Guide explains how.) The magnetic fields are strongest at the poles. That is why the iron filings are more concentrated at the poles. If we try magnets in pairs, we see complicated force fields.

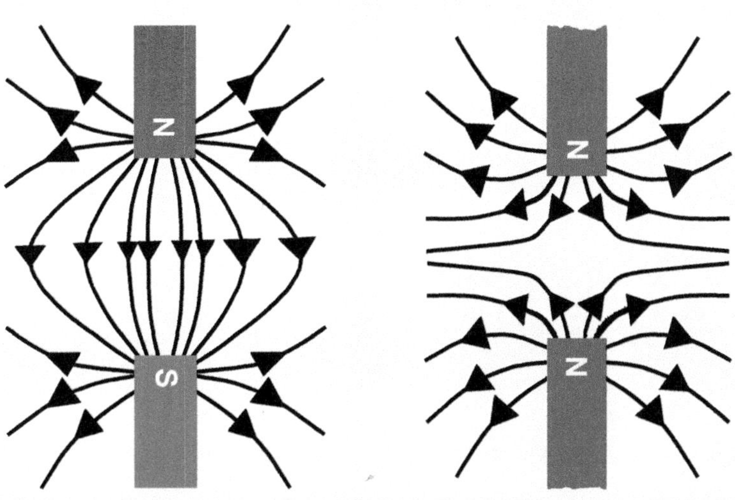

The whole Earth behaves like a magnet, which is why a compass needle points approximately north. Scientists are sure this magnetic effect is due to the interior of the Earth being one huge mass of iron.

There are three different ways of making magnets.

1. Use a coil of electric wire to set up a field – an 'electromagnet' that can be easily switched on and off.
2. Pour melted iron into a mould and place an electromagnet around it while it cools. This makes a 'permanent' magnet.
3. By contact and stroking.

Hands-on

Lay an iron nail on the table. Use one end of a bar magnet, stroke it over and over in ONE direction. See how many paper clips your nail can pick up. Now see if more stroking makes it stronger.

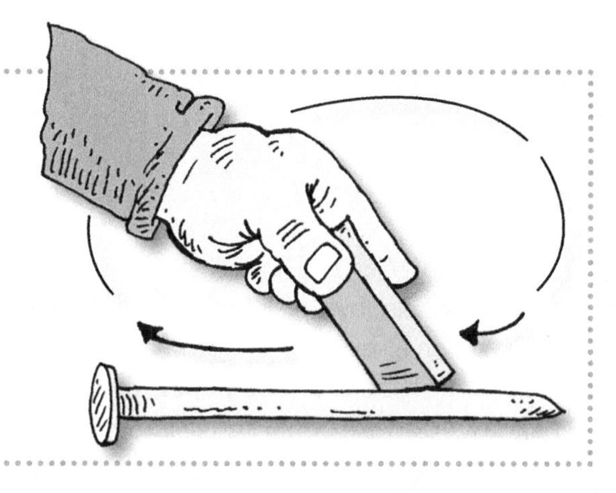

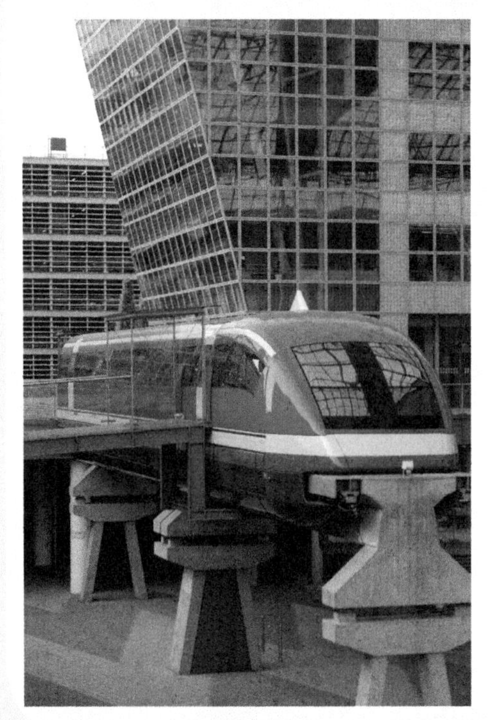

Some high speed trains use the principle of magnetic repulsion to stay above the rails. Smooth ride – but they use an awful lot of electricity!

This crane uses an electromagnet to lift scrap metal. This type of magnet uses electrical energy to operate. It is a temporary magnet, because it can be switched on and off.

Hands-on

Will magnets work through solid objects? Magnetic pull is greatest when the object is closest to the magnet. The further away, the weaker the magnetic force. Set up your experiment as this drawing shows, laying all three objects on smooth flat paper. Put the paper clip at position X. Gradually move the magnet closer until the clip is attracted and moves.

Do this a couple of times, and write down the average distance of 'first movement'. Now put your hand between the magnet and the paper clip, and get a new measurement. Then do this with one book in the way, then a piece of wood or metal. Write down all the numbers (your results). Now write down what these results tell you.

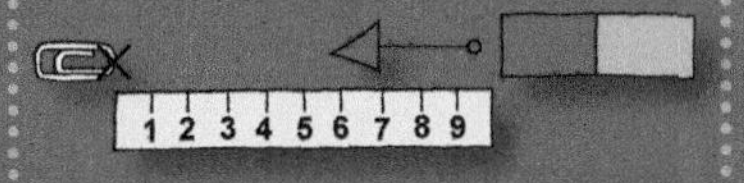

QUESTIONS

1. Answer true or false for each of the following statements.
   a. Magnetism is an invisible force.
   b. Iron and nickel are magnetic.
   c. Aluminium is magnetic.
   d. The Earth is like a giant magnet.
   e. Two similar ends of magnets will attract each other.
   f. You can make a magnet with electricity.
2. Draw the force field around an ordinary magnet.
3. Draw an arrangement of two magnets that will repel.
4. Draw an arrangement of two magnets that will attract.

Extension:

1. Find out how maglev trains work.
2. Find out how to make an electromagnet, and plan a way to measure its strength.

MELTING moments

In this chapter we will:

Learn about changes of state

Learn what happens to particles when a substance changes state

SUBSTANCES CHANGE in many different ways: iron rusts, wood burns, ice melts, water boils, fireworks explode. Some of these changes can be reversed, some can not. Some changes are quite easily reversed: we call them physical changes. For example, melted ice can be frozen solid again. Some changes are not easily reversed: we call them chemical changes. For example, you can't uncook an egg, and you can't unburn wood.

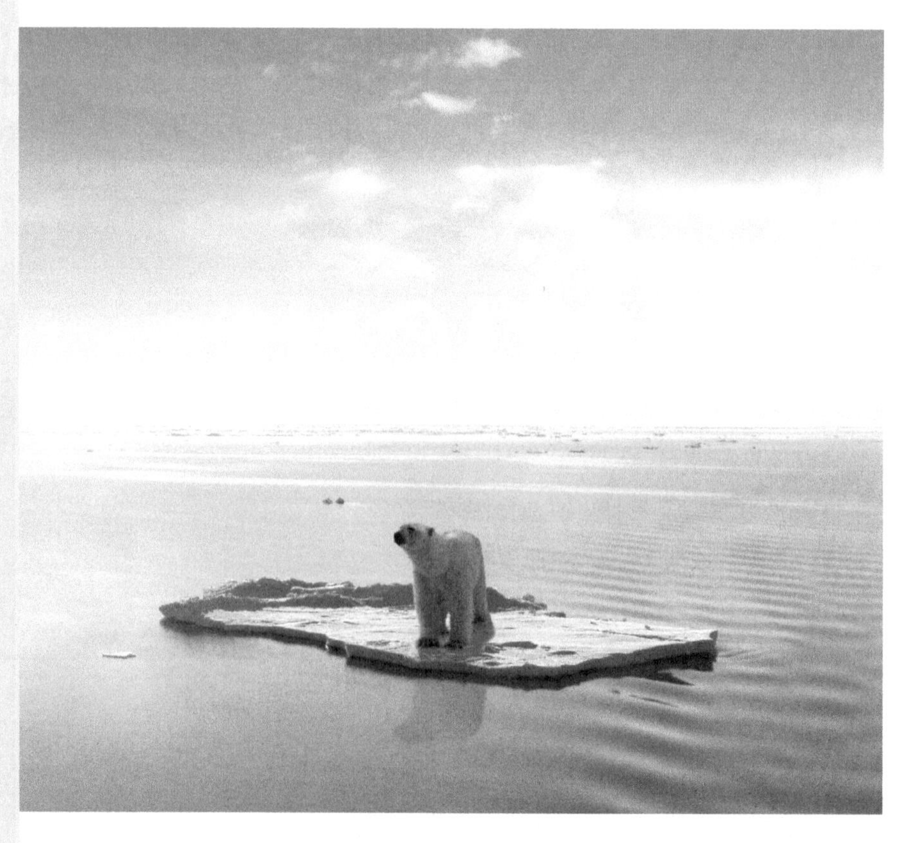

Any change of state is a physical change. What's actually going on with the particles? The hands-on with candle wax will give you more ideas.

Cut some candle wax into small pieces. Put the pieces in a glass test tube until it is about one third full. Using something to hold it, place the test tube over a small flame and heat it gently until the wax melts. Hold it upright and carefully mark the liquid level. Now leave it upright to cool for 5 minutes.

- Describe what you saw.
- What happens to the level of the wax when it changed back to a solid?
- Are you able to melt the wax and go through the process again?

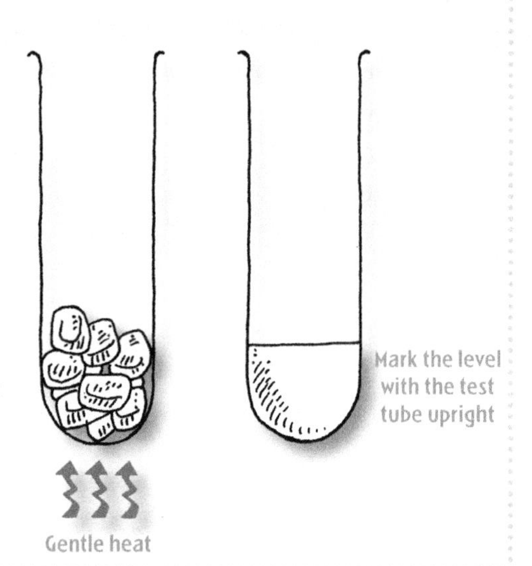

What is going on inside wax when it changes from a solid to a liquid, then back to a solid again? It's those invisible particles again. In a solid they more or less stay in place. When the solid is heated the particles start to vibrate faster and rush around. Eventually they break away from their neighbours and have no fixed places, so the solid changes into a liquid. When a liquid is heated even more the particles move far apart, and it becomes a gas.

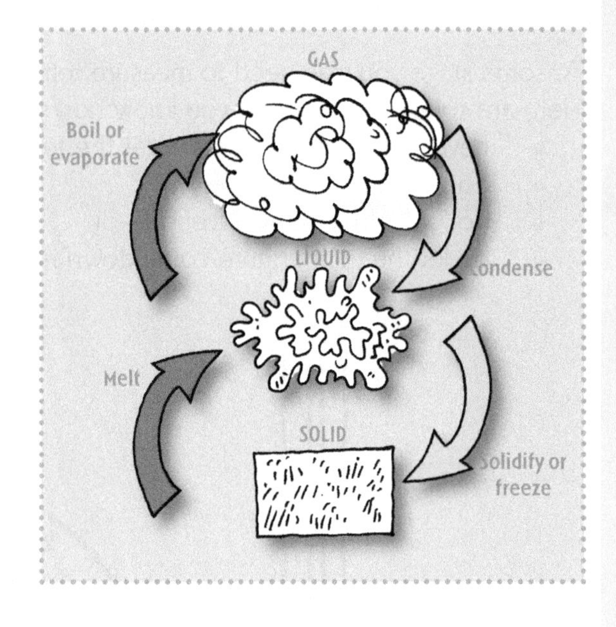

Evaporating and boiling

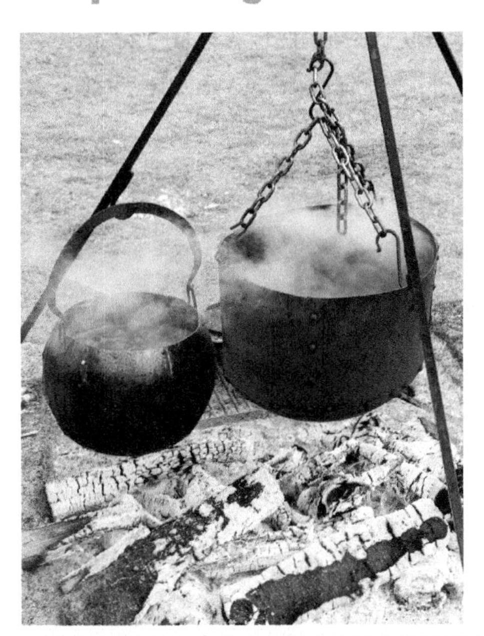

When washing dries on a line, the water evaporates – this means it changes from a liquid to a gas. If a can of water is placed on a fire and the water gets to 100°C then the temperature won't rise any more. This is the boiling point. The water evaporates (boils) very fast and goes into the air as steam. The water doesn't get any hotter because all the heat is being used to make water change state from liquid to gas. Water particles in the gas state (steam) have more energy and move much faster than in liquids.

What about the opposite? We call this condensing. When steam cools down to under 100°C the water particles join together in droplets of liquid water.

Many materials have definite boiling and melting points.

Pure water freezes at	0°C
Ice melts at	0°C
Pure water boils at	100°C
Copper melts at	1084°C
Iron melts at	1538°C

The C stands for Celsius or for centigrade – it makes no difference.

At some stage you may need to measure temperatures.
Here are some basic rules so you know how to use a thermometer properly:

- never put it near a flame or let the temperature get over 105°C
- don't put it in your mouth
- do look at the liquid level side-on
- do let the temperature come down naturally: don't shake the thermometer.

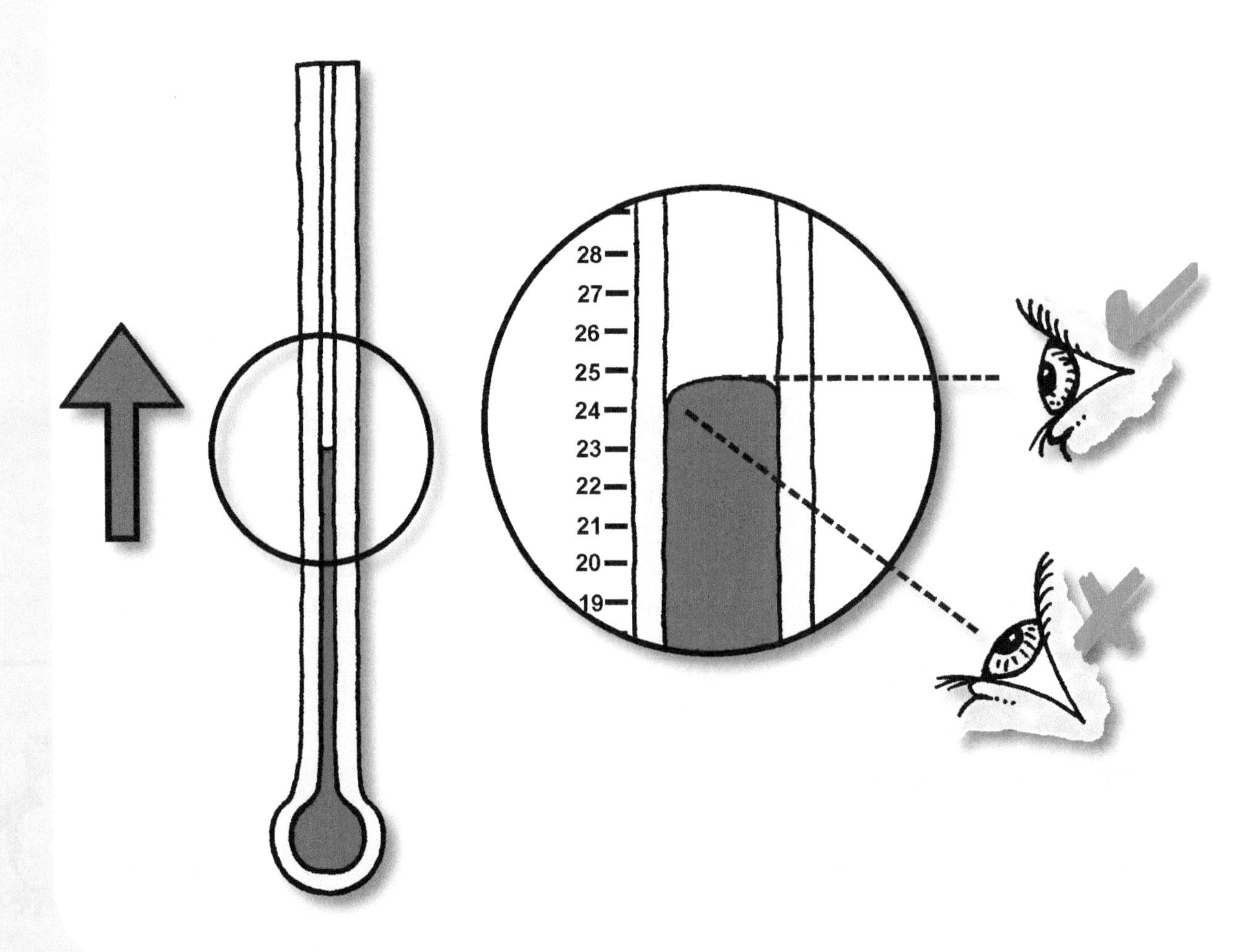

Hands-on

When water changes from one state to another, no new substance is formed. It is still water, but just in a different state: solid (ice), liquid, and gas (steam or water vapour).

For safety, this activity is best done with a teacher's help. It takes three people to do it properly: one to watch the thermometer, one to watch the time, one to write things down. Start with a container like a beaker, and pour in an exact measured amount of water. Mark its level and take its temperature. Put the beaker on a hot plate and carefully measure the temperature every minute. Keep going until the water has been boiling for at least a few minutes.

Write down your results in a table like this. Let the water cool.

Time in minutes	1	2	3	4	5
Temperature °C					

- What did the temperature rise to?
- What happened to the temperature after this?
- What else did you notice?
- What happened to the water level?
- Where do you think the lost water has gone to?

QUESTIONS

1. What do we called the change of state when we let steam change back into water?
2. What do we call the change of state when we make water change back into ice?
3. Why is it possible to boil water in an electric plastic kettle, without the plastic melting?
4. Explain why the level of the wax went down when it changed from a liquid to a solid (page 45). Use the word 'particles' when you explain this.
5. What happens to water when it is cooled to 0°C?
6. What happens to copper when it is heated to 1084°C?
7. Draw 10 particles of water to show how they move at 20°C, -5°C, and 107°C.

Extension:

Find out the boiling, melting and condensing points of at least ten solids, liquids and gases.

taken apart

In this chapter we will:

- Learn that a mixture is not the same as a compound
- Learn about different examples of mixtures
- Learn about different ways of separating mixtures

WHY IS it easy to get salt out of sea water, but hard to get hydrogen out of sea water? The reason: there is a big difference between a mixture and a compound.

A mixture is simply two or more things in the same place at the same time, but not joined together. Just like fruit mixed in the same bowl: the apples are still apples and the oranges are still oranges etc, and it's quite easy to separate them. Another example: sea water is a mixture or salt and water, and it's quite easy to separate the salt from the water.

A compound is different from a mixture. When two different kinds of atoms join together they form a compound, and a compound is not easily broken apart. For example, salt is a combination of sodium and chlorine. It's very difficult to separate these two. It is also difficult to separate hydrogen and oxygen when they join to make water.

Look at this table for some examples.

Mixture its parts are fairly easily separated	Compound atoms are joined, not easily separated
Air: a mix of many gases	Water: H and O atoms
Sand: different kinds of grain	Petrol: C and H atoms
Sea water: water and several kinds of salt	Sugar: C and H and O atoms

Hands-on

Mix together some rice, sand, and small pebbles. Use an ordinary kitchen sieve to separate these. You may need a second sieve to separate the mixture completely into three different piles. What kind of second sieve would you need?

In a science room, we often use a very fine kind of micro-sieve called filter paper. This lets atoms and very small particles go through, but holds back particles bigger than about 1/100mm diameter.

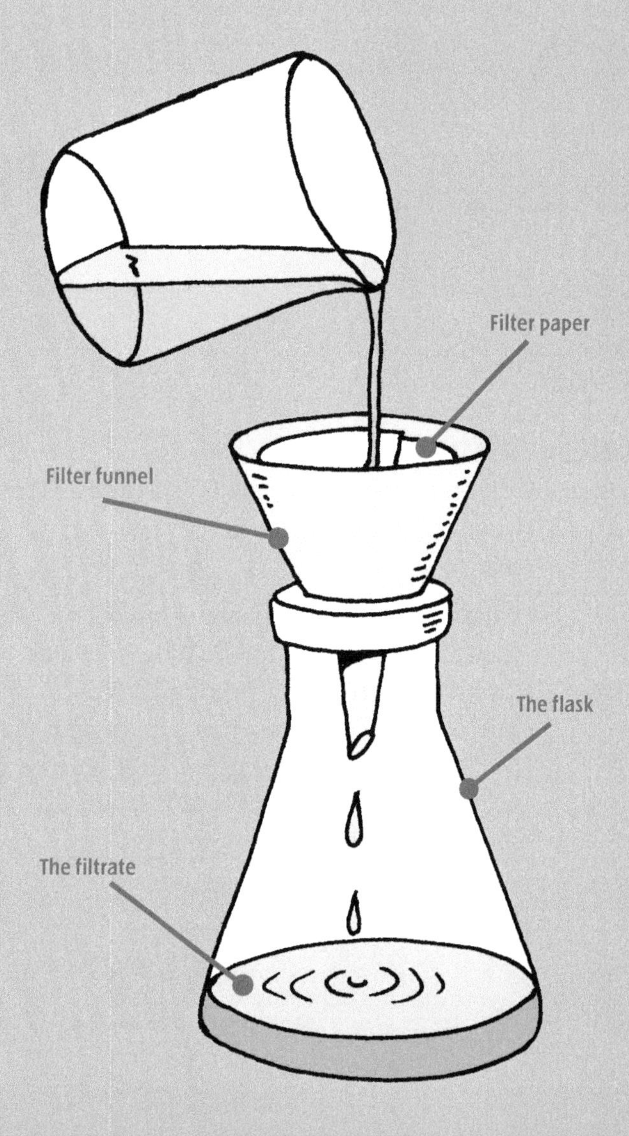

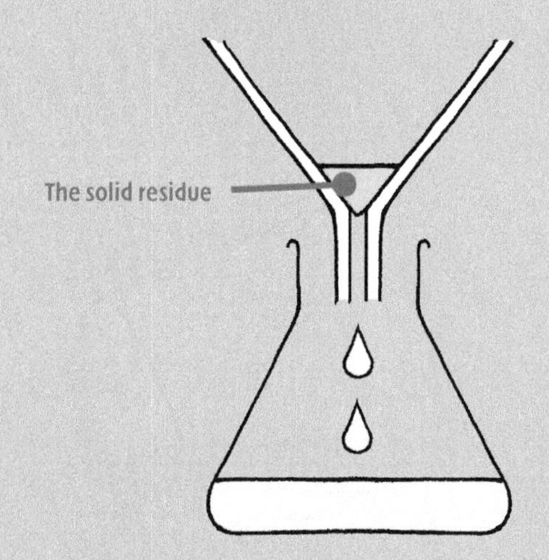

Some different kinds of mixture

Sea water is 97% water, plus many different kinds of salt. There is even a tiny amount of gold! What do you think will happen when you put sea water through a filter? Will the salt stay behind on the filter paper or not? You could try this out.

Try looking at some beach sand down a microscope!

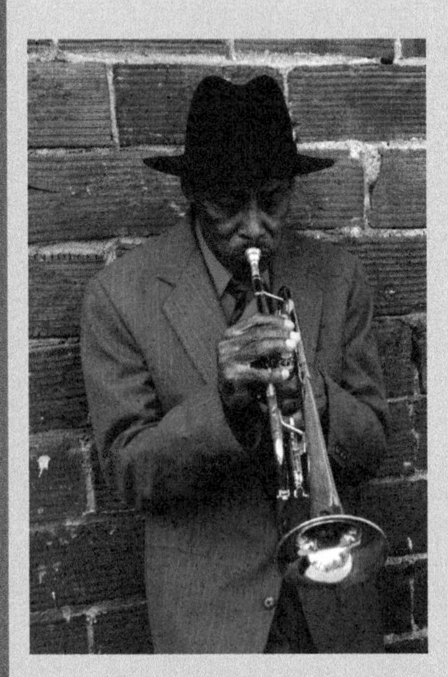

Alloy mixtures

'Alloy' is the word for a mixture made of two or more kinds of metal. Some examples:

- Brass is a mixture of copper and zinc.
- Bronze is a mixture of copper and tin.
- Stainless steel is a mixture of iron and nickel and chrome.

Separating mixtures

There are many different ways of separating mixtures. Here are six you could try out.

Sieve	if the particles are medium to big.
Filter	if the particles are medium to small.
Pour off	such as for a layer of oil floating on water.
Evaporate	if you just want to get rid of a liquid.
Magnet	works if part of the mix has got iron in it.
Dissolve	if one part of the mixture will dissolve in water and one won't, this can be a first step to separating them.

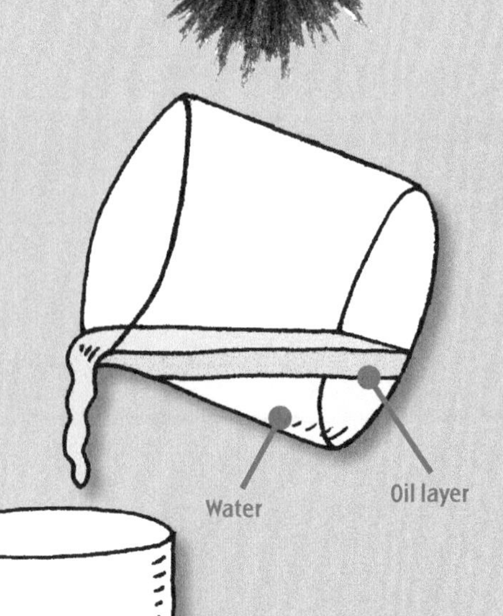

More about compounds

When atoms of different kinds combine to form a compound, they always do so in particular exact numbers. For example, when H combines with O it is always two hydrogen and one oxygen, which is why we write H_2O. A water molecule never has one or three or four H atoms. A mix is different, because it can be in any amounts. You could mix one spoon of salt into a litre of water, or five spoons.

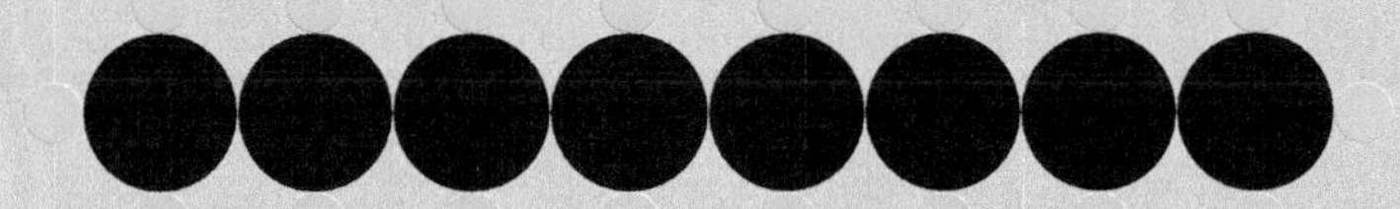

Some molecules are big! This is a molecule of octane, a substance in petrol. Octane has 8 carbon atoms and 18 hydrogen atoms.

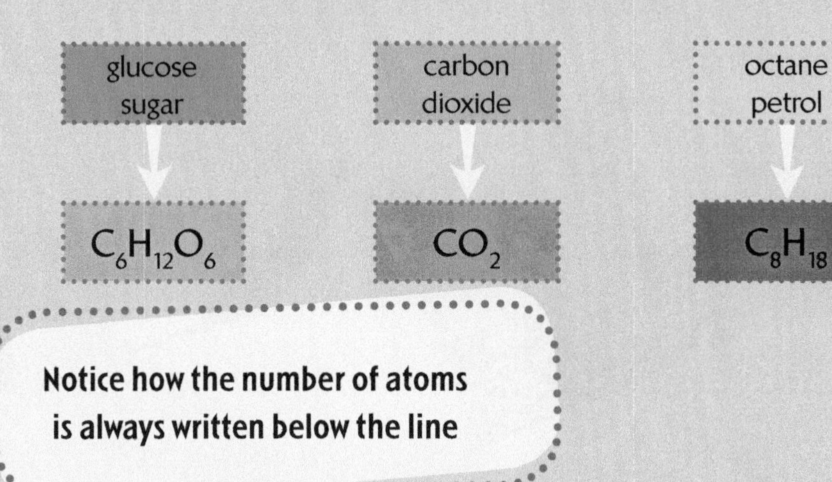

Hands-on

You will be given a mixture that is exactly 10 grams kitchen salt and 10 grams beach sand. Your aim is to separate the mixture back into 10 grams of dry sand and 10 grams of dry salt – without losing any down the drain or onto the floor!

First, make a plan using some of the separating methods you have just read about (page 50).

Next: do it!

At the end, weigh the sand and salt again, and answer the following questions.

- Did you have exactly 10g of each? If not, suggest what went wrong and what you could do better next time.
- Did the sand taste a little bit salty? If so, suggest how you could prevent this happening next time.
- How did you dry the salt and sand? How could you do this a bit faster next time?

QUESTIONS

1. Name two examples of mixtures.
2. Name three examples of chemical compounds.
3. Name two differences between a mixture and a compound.
4. What is meant by an alloy? Give two examples, and say what is in each.
5. List five ways of separating mixtures.
6. Describe how you would separate a mixture of oil and water.
7. Describe how you would get salt out of sea water.
8. Describe how you would separate sand from iron-sand.
9. Name two liquids that easily mix together.
10. Name two liquids that won't mix together.
11. How many atoms are there in a molecule of glucose sugar?

Extension:

Mix up sand, sugar, cooking oil, and iron-sand. Now make a plan to separate this messy mixture again. List all the equipment you would need and all the steps you would take. Now see if you can do it – but clean up afterwards!

SORTING things out

In this chapter we will:

- Learn that each substance has its own special properties
- Learn different ways of describing substances and objects
- Learn about different kinds of atoms
- Learn about molecules

HAVE A look at the photographs below, and decide what substance each is made of. Now choose words to describe the properties of each substance. 'Properties' means 'features that are special to it'.

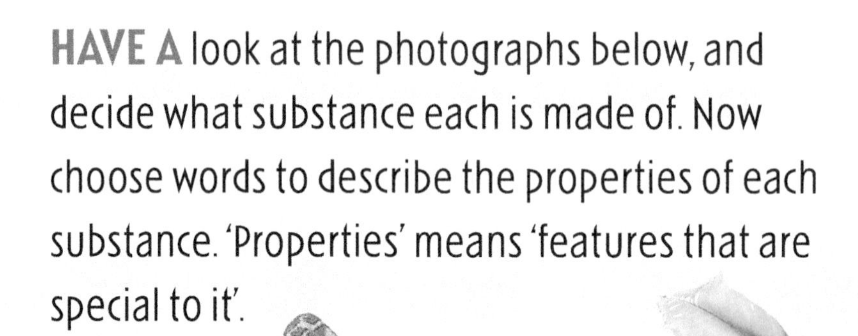

When you think about it, each substance has many different properties. Here are eight properties, but you could probably think of more.

1 Is it solid, liquid, or gas?
2 If it's solid – is it flexible or does it have a fixed shape?
3 Dense (heavy), or light (floats in water or air)?
4 Does it dissolve in water?
5 Does it dissolve in acid?
6 Does it conduct electricity?
7 Does it burn?
8 Does it smell?

You can do this as a thought experiment, or you can actually try it out and see with your own eyes. First, start with a substance of some kind, like a piece of tyre rubber. For example, you could say this about tyre rubber: 'solid, very flexible, sinks in water, burns slowly, smells rubbery, not sure if it dissolves in acid.'

- Now describe the properties of ice. The list of eight will help you, but you could add other properties.
- Now do the same for aluminium foil.

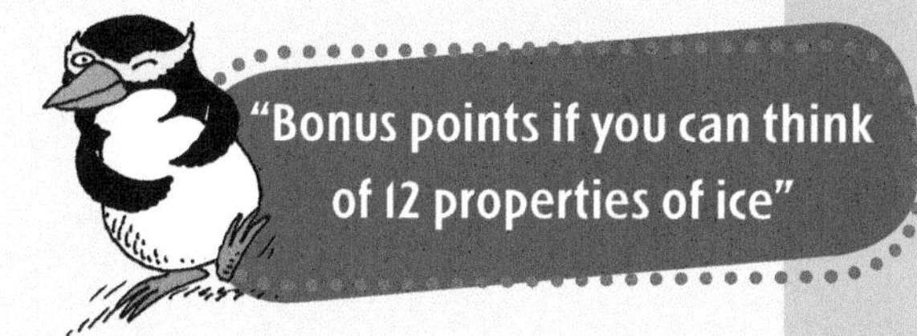

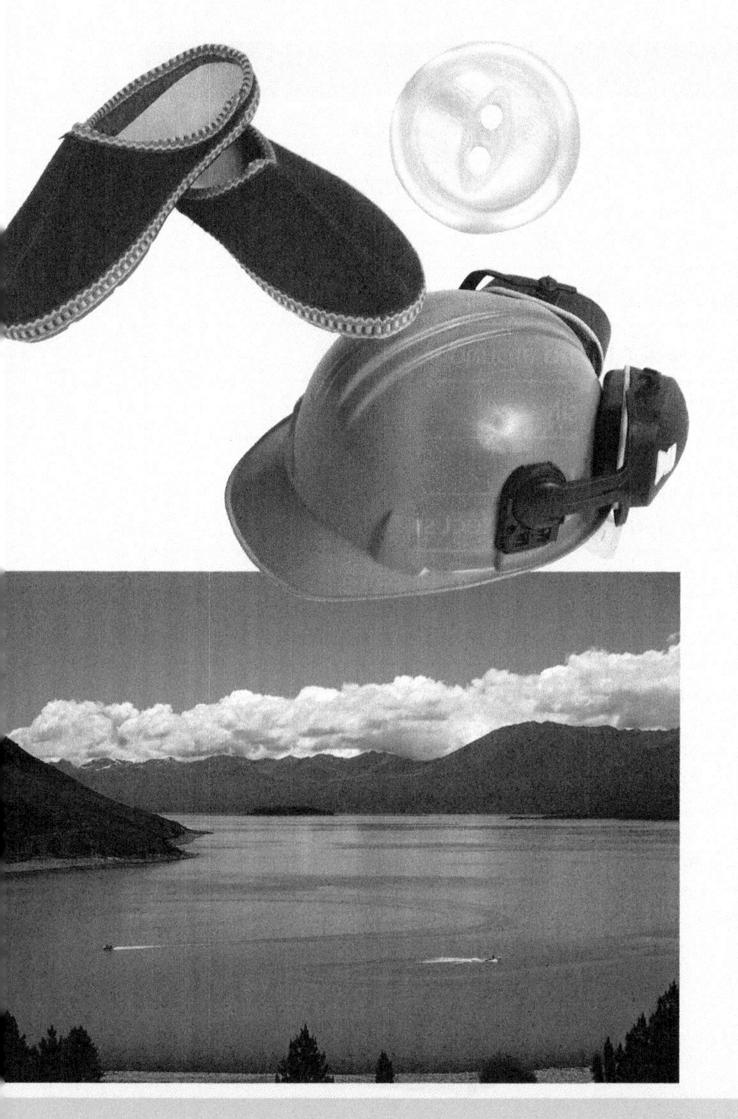

In the hands-on test on page 55 we use three metals: lead, aluminium, iron. These three are what we call 'elements', and the special properties of each are the result of the kind of atoms it has.

What is an 'element' ? The word has more than one meaning. In olden times the word 'element' meant earth, air, fire, and water. It can also mean the hot plate on a stove. In science the word 'element' is used for a pure substance made of one kind of atom. We know of 103 different kinds of atom, and between them they make up every kind of material in the world.

The good news is you don't have to learn 103 names, but it helps to know some of the common kinds when you are sorting things out. Elements are the basic building blocks of every material thing in the universe, from your fingernails to stars. Here are a few element names: hydrogen, carbon, oxygen, sulfur, gold.

To save you having to write out these names, science has a shorthand system of 'symbols' with one or two letters standing for each element. If there are two letters, we always write the second letter small. Do your best to learn the 15 names and symbols in this table.

Name	Symbol	Some facts
Hydrogen	H	The smallest atom. A very explosive gas.
Carbon	C	The black solid in pencils. Can also form diamonds.
Nitrogen	N	Makes up 78% of the air.
Oxygen	O	Makes up 20% of the air.
Sulfur	S	A yellow powder found around volcanoes.
Magnesium	Mg	Light metal, burns easily.
Aluminium	Al	A very light metal.
Chlorine	Cl	Poisonous gas, used to disinfect swimming pools.
Phosphorous	P	Very dangerous in its pure element form.
Iron	Fe	Hard metal. Its symbol is from the French word 'fer'.
Zinc	Zn	Metal used for rust-proofing.
Copper	Cu	A metal that conducts electricity very well.
Lead	Pb	Its symbol is from the Latin word 'plumbum'.
Gold	Au	Its symbol is from the Latin word 'aurum'.
Uranium	U	A big atom. Can be radioactive.

"Plum bum?
Are you kidding?"

Some atoms are real loners, others like to team up. Gold is a loner; it never teams up. Oxygen atoms like to go around in pairs, so we write O_2 for short. Two oxygen atoms can team up with carbon to make carbon dioxide, so we write CO_2 for short.

Any substance which has different kinds of elements joined together is called a 'compound'. Often, compounds look very different to the elements that make them up. For example, copper sulfate (blue crystals) is a compound of three elements: copper, sulfur and oxygen.

Copper

Sulfur

Copper sulfate crystals

Groups of atoms are called 'molecules', whether there are two in a group, or 200. Often we just use the word 'particle' to cover all kinds of atoms and molecules.

Hands-on

Your aim: to compare the properties of three different kinds of metal: lead, aluminium foil, iron. (Or you could use three different kinds of plastic.)

Here are three properties that are easily tested: flexibility, colour, shininess.

Plan your own simple tests for each and add a couple more if you can. Make a table with a column for each material and a row for each property.

Write down each result as you see it.

QUESTIONS

1. What do we mean by the 'property' of a substance?
2. Name four or more properties of diamonds.
3. Name four or more properties of petrol.
4. Name two or three properties of aluminium that make it useful for boat-building.
5. Explain what we mean by an 'element' in science.
6. Write down in one column the names of 15 elements named in this chapter. Now without looking, write the symbols of each next to its name. Now check the book and write in the ones you missed out.
7. Explain the difference between 'atom' and 'molecule'.
8. Draw a molecule of O_2 and a molecule of CO_2.

Extension:

Find more facts about any ten elements.

SOUNDING waves

In this chapter we will:

- Learn how sound travels in air
- Learn about what we mean by frequency (pitch)
- Learn what we mean by wavelength

VOICES, MUSIC, cars, wind, animals: our lives are surrounded by sound. How does sound actually get through the air and inside your brain?

Let's start with this guitar playing the note 'A', and imagine it in slow motion. As the 'A' string is plucked it vibrates back and forwards at high-speed. Each time it moves forwards the string pushes on air particles which push on neighbouring air particles. The result is that little waves of air pressure spread outwards, like ripples in a pool of still water.

These waves travel through air at the speed of sound: about 320 metres every second. Sound travels faster under water and slower in high-altitude thin air. Sound waves move very slowly compared to the speed of light, which zips along at 300 million metres a second.

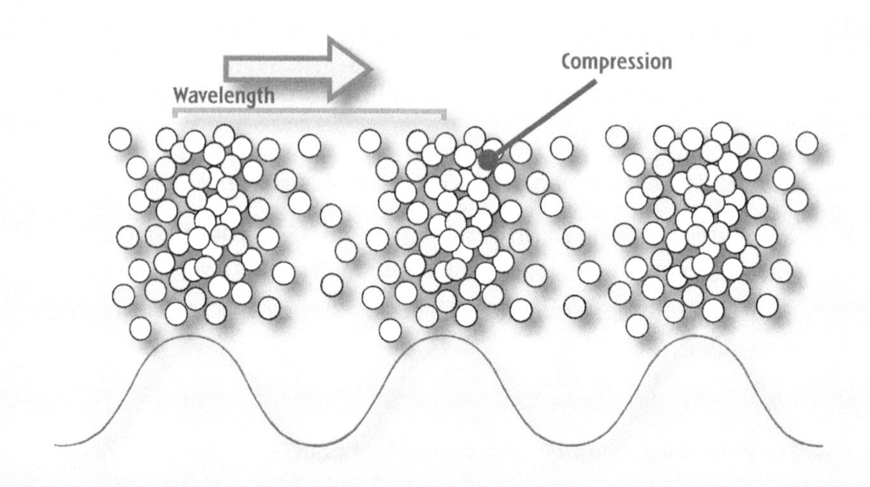

The 'frequency' of sound is how many waves there are each second. For example, the note 'middle A' has a frequency of exactly 256 waves a second. Frequency is also known as 'pitch'.

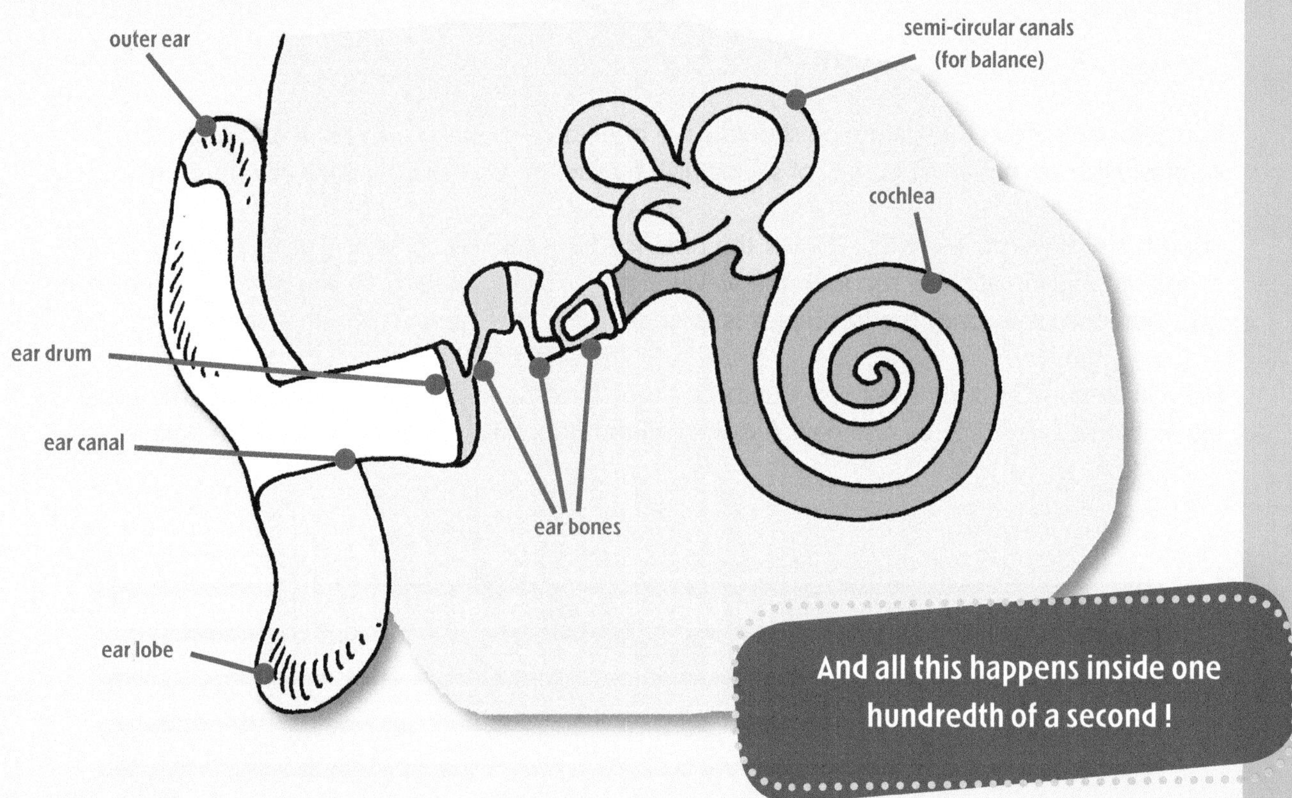

And all this happens inside one hundredth of a second !

The outside part of the ear collects sound waves and funnels them into the ear canal. At the end of the canal is the eardrum, a thin piece of skin that vibrates when sound waves hit it. The eardrum makes three tiny ear bones vibrate. The vibrations are then passed to the cochlea – a system of tiny tubes inside the ear. The liquid in these tubes triggers sensitive nerve cells. Nerve cells pass the message on to the brain which interprets it as a sound.

Stretch a rubber band very tightly over a book and place two pencils underneath the rubber band as the picture shows. Gently pluck the rubber band and listen for sound.

Move the pencils closer together. Keep moving the pencils closer and plucking the rubber band in between them each time. What happens to the sound as the pencils get closer together?

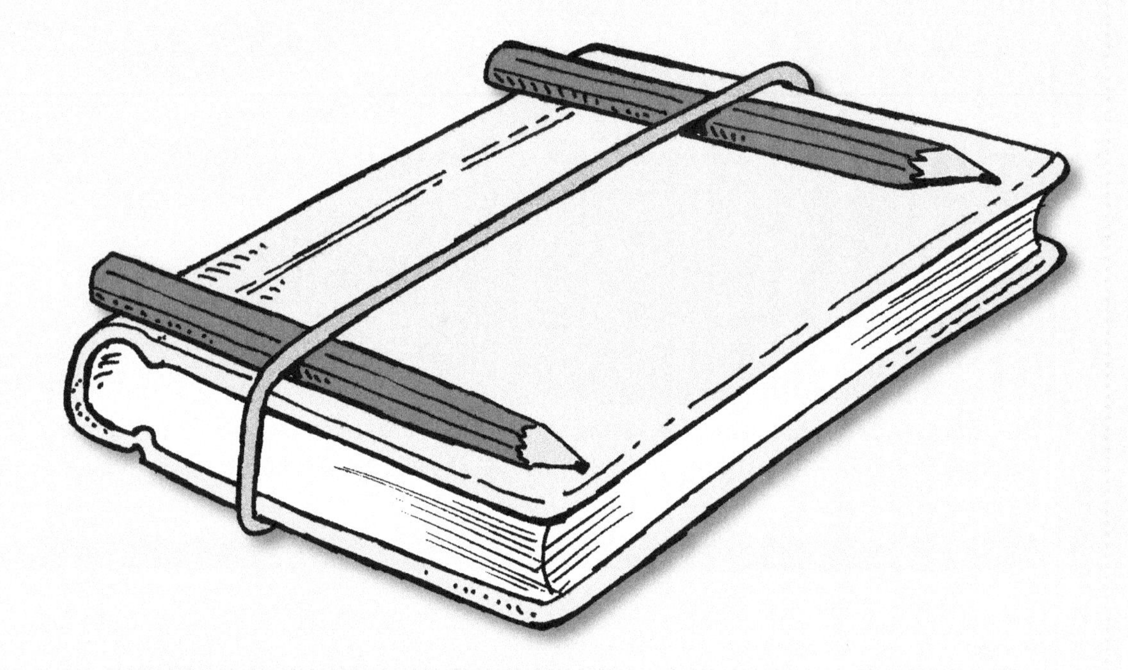

Your rubber band in the Hands-on should have been able to make high pitch and low pitch sounds. High pitch is another way of saying high frequency. Low pitch means low frequency.

Sounds also have wavelength, which is the distance from one compression wave to the next. At middle A (256 vibrations a second), the wavelength is 75 centimetres. At the next A up the scale (512 vibrations a second) the frequency is twice as much and the wavelength is half.

Musical composers don't use these numbers – they have their own system of writing. The usual musical scale is ABCDEFG. The next A is the eighth note, which is why it is called an octave.

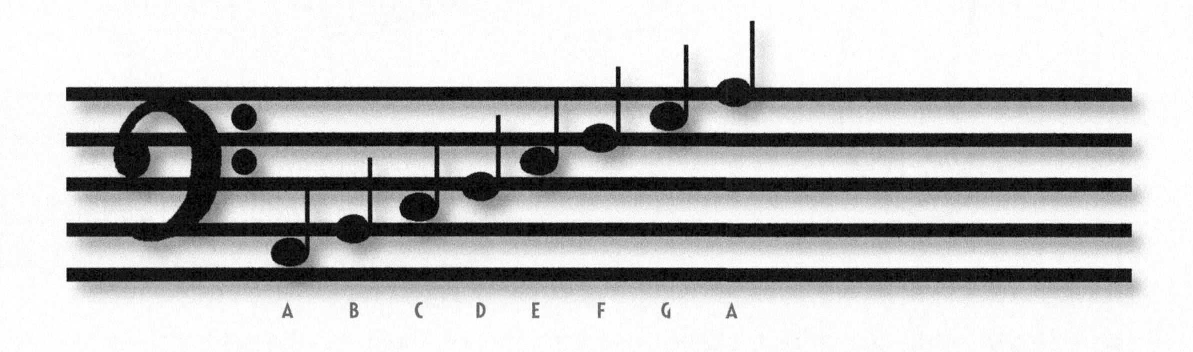

Hands-on

Making music. Get a few same-shape glass bottles and fill them with water to varying levels. Tap each bottle with a pencil and compare the pitch of the sound coming from each one, then arrange the bottles from highest to lowest pitch.

Which bottle gives the highest pitch? And the lowest? With a bit of care you can arrange a musical scale and play simple tunes.

QUESTIONS

1. Describe in 'steps' how sound waves travel from a guitar to your cochlea.
2. Draw a diagram to show what 'wavelength' means.
3. Name the part of the inner ear that changes vibrations to nerve signals.
4. Give another word for frequency.
5. Explain in five to ten words what frequency actually means.
6. Give the wavelength of the note 'middle A'.
7. Work out the wavelength of the 'A' note, one octave higher than 'middle A'.

Extension:

Find out the speed of sound in air, water, steel. What makes sound travel at different speeds?

UP IN the air

In this chapter we will:

- Learn that air is a mixture of many different kinds of gas
- Learn that living things depend on two of these gases in particular
- Learn how to test for carbon dioxide

DO YOU know that if the air was pure oxygen, everything would catch fire? Let's have a look at what air is actually made of.

What's air made out of? Nitrogen (78%) and oxygen (20%) are the two most common gases in the air. Some of the less common ones are helium, argon, water vapour and carbon dioxide. The amounts are not much different from one part of the world to another because winds mix them all up.

How high does the air go? Although air goes up to more than 1000 kilometres above ground level, most is contained in the bottom 10 kilometres. Higher than this, air is too thin and too cold for living things to survive. If you imagine the earth scaled down to the size of a basketball, 10 kilometres of air would be a paper-thin layer.

Are there atoms in the air? Yes, of course. All substances have atoms.

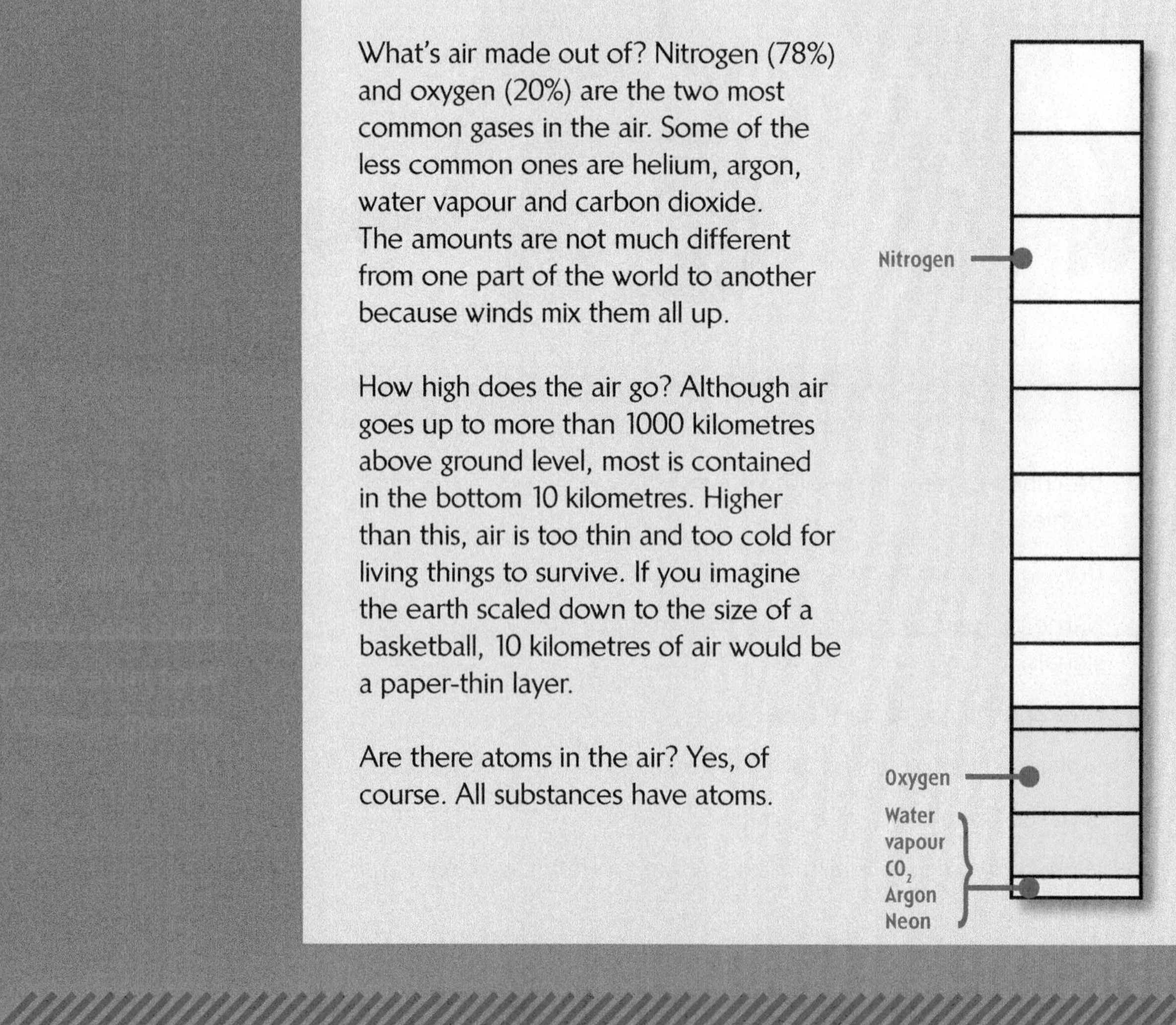

Atom picture	Name of this gas	Symbols	
	Argon	Ar	These are **elements** (one single kind of atom in each, even though some are in pairs)
	Nitrogen	N_2	
	Oxygen	O_2	
	Carbon dioxide	CO_2	These are **compounds** (two or more different kinds of atom joined together)
	Water	H_2O	

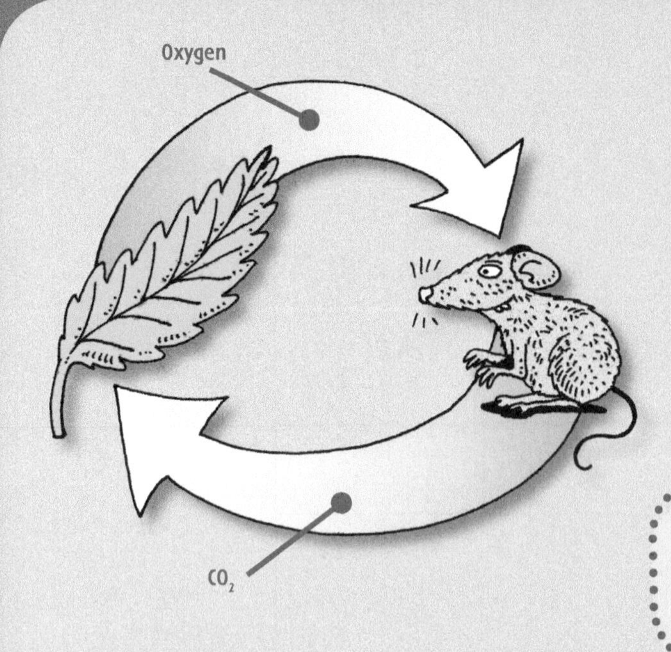

Air is being used up all the time and replaced all the time. Every day millions of tonnes of oxygen is used up by animals, humans, cars, planes. Every day almost exactly the same amount of new oxygen is put back into the air by plants. In fact every atom of oxygen in every breath you take has been put there by green plants. As well as giving off oxygen, green plants take in carbon dioxide and use it to make sugars.

So a leaf is a solar-powered sugar factory?

This shows what we breathe in and out. The oxygen is used to burn up sugars, which provide energy for our muscles. The air that you breathe out has less oxygen and more carbon dioxide than normal air. In a few days, winds will have spread your breath over most of the world. Sooner or later, all the carbon dioxide you breathe out is used up by green plants. So you end up spread over a big area!

What else are we doing to the air? Cars, planes, fires and factories all pour huge amounts of carbon dioxide into the air: about 15 million tonnes every day worldwide. Carbon dioxide is not poisonous, but it acts like a blanket to keep heat in. It is thought that the increase in carbon dioxide in the atmosphere has become a big cause of increased global warming and climate change.

Does air weigh anything? If all the air is sucked out of a strong 10 litre container, it will lose about 8 grams. So 1000 litres (one cubic metre) weighs about 800 grams. This means that the air in your classroom weighs about 120 kg.

It's all this heavy air !

'Lime water' is a clear liquid that turns a milky colour when it meets CO_2. (Your teacher will supply some.)

- Put three lots of 20 ml in three clean flasks, A, B and C.
- Use a straw to blow your breath through the lime water in A.
- Cover the top of B and shake it up with the air that is already inside.
- Use a fourth flask to capture the fumes above a burning candle flame, seal it to keep the fumes in until they have cooled, then pour in the lime water from flask C and shake it up like you did with B.
- Time how long it takes to get a 'milky' result with A, B and C. Write down your results. What do the results tell you?

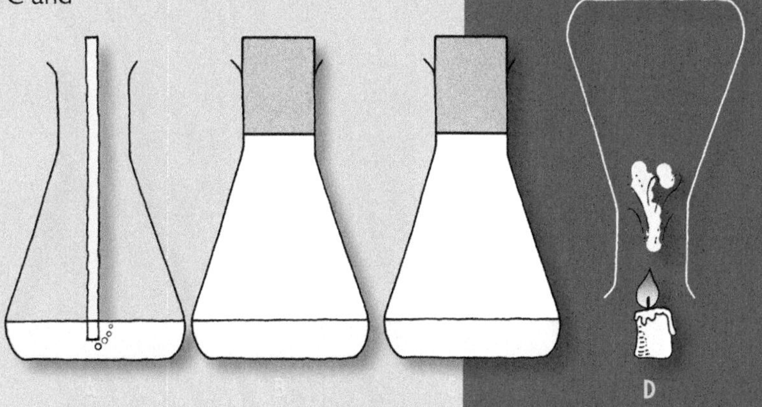

QUESTIONS

1. Name the most common two gases in the air.
2. Name four less common gases in the air.
3. Approximately how much of the gas you breathe out is carbon dioxide?
4. Where does the oxygen in the air all come from?
5. Name three things that take oxygen out of the air.
6. Count how many atoms are in one molecule of each of these: oxygen, carbon dioxide, water.
7. Describe what happens over the next few weeks to the carbon dioxide in your next breath. You will have to use your imagination for this, but realise that the wind will blow it far away, and that plants will sooner or later use all of it, and that animals are likely to eat the plants, and then...

Extension:

Ozone

Find out what ozone gas is, and where it is, and how it is naturally made. Why is it so important to life on Earth? What is causing it to disappear? What can we do about this?

CLEAR water

In this chapter we will:

- Learn about the water cycle
- Learn about farm fertilisers
- Learn about the effect of fertiliser on lakes and rivers

NEW ZEALAND has some of the clearest air and cleanest waters in the world. Some rivers and lakes are so clear that you can see down through ten metres of water. But the situation in others is not so good.

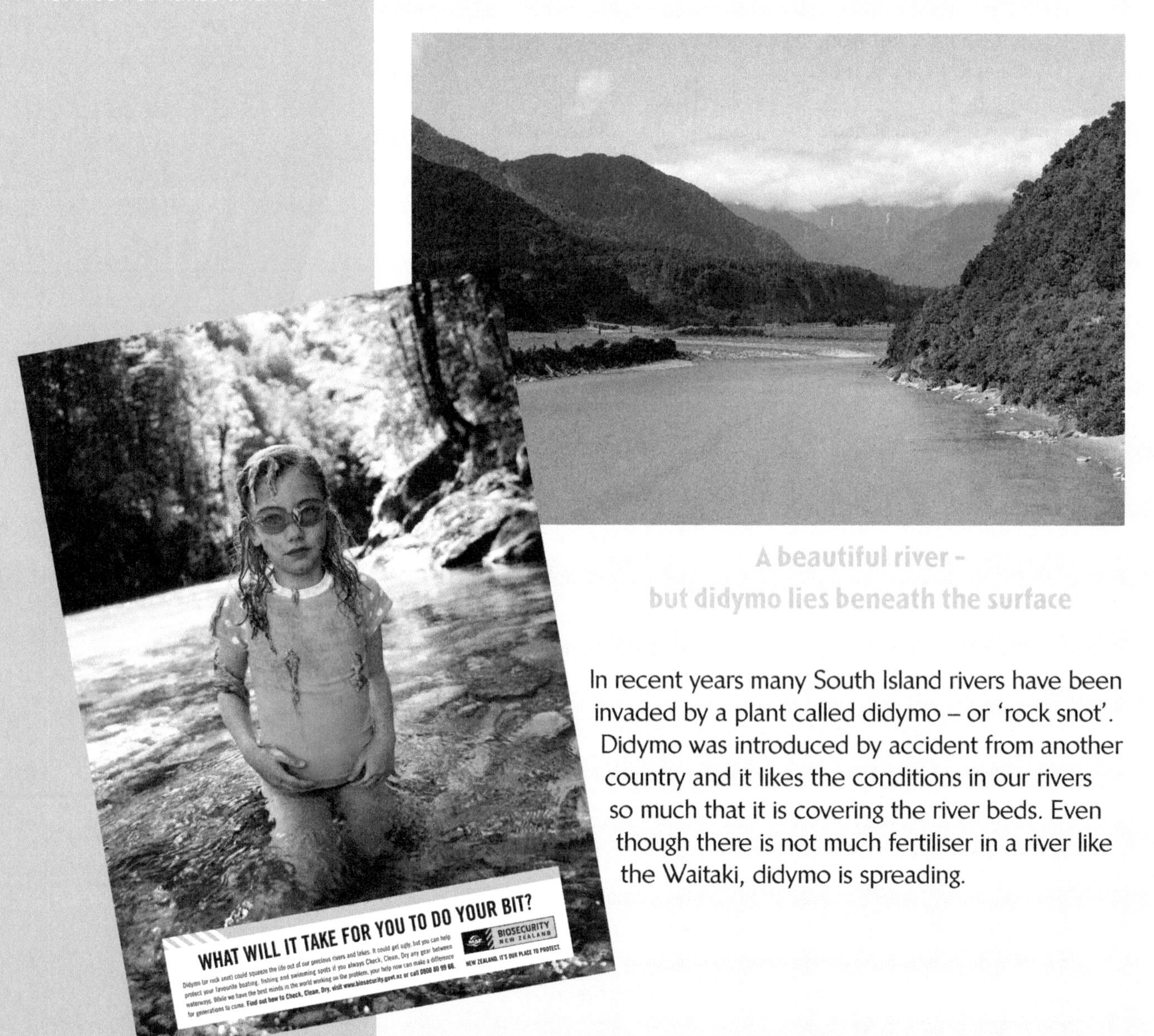

A beautiful river -
but didymo lies beneath the surface

In recent years many South Island rivers have been invaded by a plant called didymo – or 'rock snot'. Didymo was introduced by accident from another country and it likes the conditions in our rivers so much that it is covering the river beds. Even though there is not much fertiliser in a river like the Waitaki, didymo is spreading.

This drawing shows how water is recycled in nature, and why we never come to the end of water. A full cycle takes months or years, and every drop of water in the world and in your body has been through many cycles already.

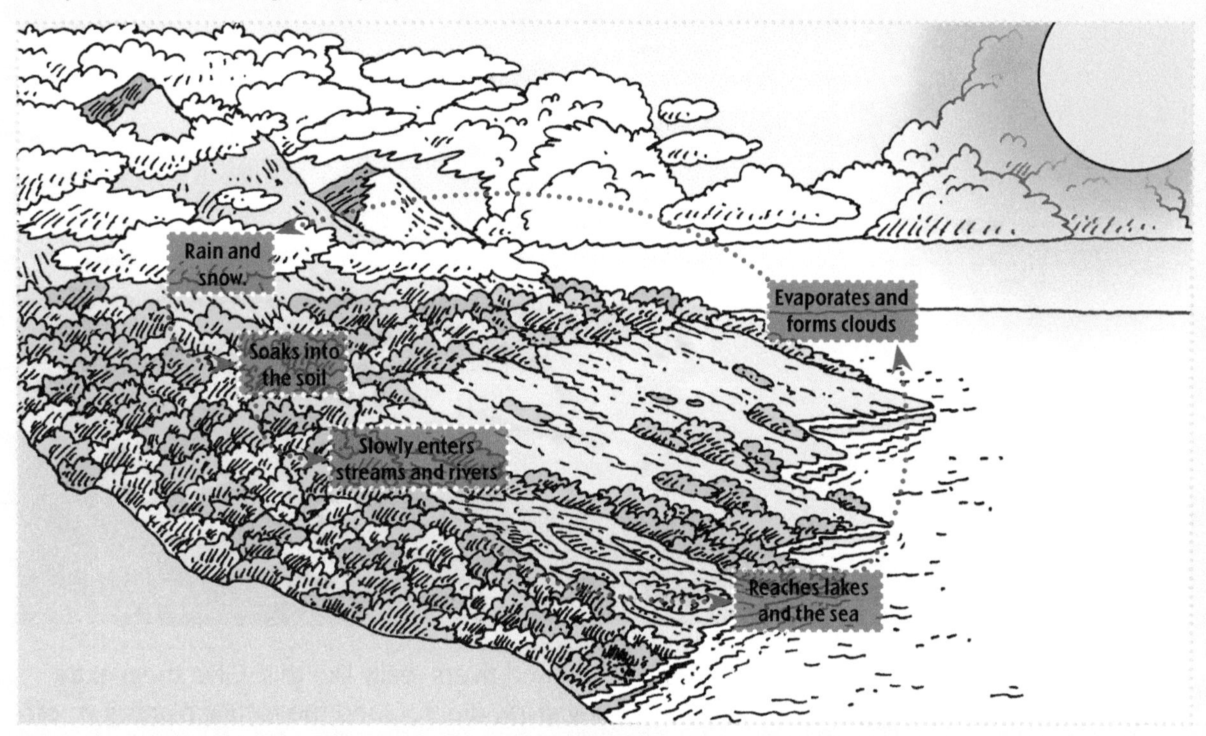

Maori people have always valued clean water, and they have tapu and tikanga rules on how to use water supplies. For example, human waste should not be allowed near water.

tapu:	sacred and forbidden
tikanga:	traditional and proper
kaitiaki:	responsible care of food supplies

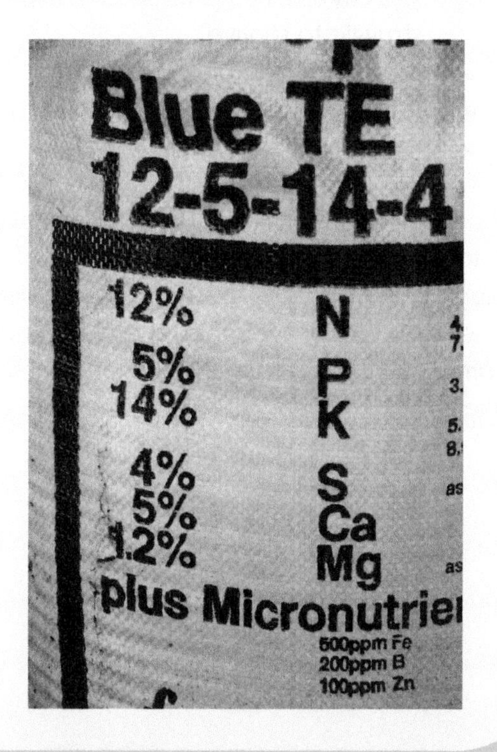

Most New Zealand farms are efficient and well-run. New Zealand's dairy cows can produce more milk from one hectare of grass than cows from anywhere else in the world can. To keep the grass growing as fast as possible, farmers use plenty of fertiliser, especially fertiliser that has a high proportion of nitrogen (N), phosphorus (P) and potassium (K).

Not all fertiliser gets absorbed into the grass. Much of it gets washed by the rain into streams and rivers then into lakes. Some cow poo gets washed off farmland in the same way.

The algae and other plants that naturally live in lakes and rivers really like this. Give them extra fertiliser, and they start to grow fast. Very fast. They grow, die, rot, and the rotting plants give off nitrogen (N), phosphorus (P) and potassium (K) back into the water again.

This over-growth of water plants has a big name: 'eutrophication' (say *you-trofy-cayshun*). It means 'well-feeding'.

Eutrophication can be a big problem, because

- the plants die and rot and stink
- rotting takes oxygen out of the water
- less oxygen means fish and smaller life struggle to survive
- it can look horrible
- it tangles boats and fishing lines
- it can make swimming dangerous.

The whole situation is a big problem in Lake Rotorua and some other lakes nearby. Lake Taupo is starting to go the same way. In both cases the cause is the same: chemical runoff from human and farm activity is getting into the lake.

The aim is to see the effects of fertiliser on water plants.

Start with three clean clear 1.5 litre bottles. Put exactly 1 litre of tap water in each. Add a small pinch of garden fertiliser to bottle A, a teaspoon of fertiliser to bottle B. Leave C as it is. We call C the 'control' for the experiment.

- Place A, B and C in a place where they get good sunlight for several hours every day.
- Now put a small amount of algae or water plant in each, and a 10 cm piece of oxygen weed as well.
- Think about how you will make conditions in A, B and C fair and the same in every other way.
- Now watch the plants every week for three months at least.
- Once a month draw and describe the situation in each bottle.

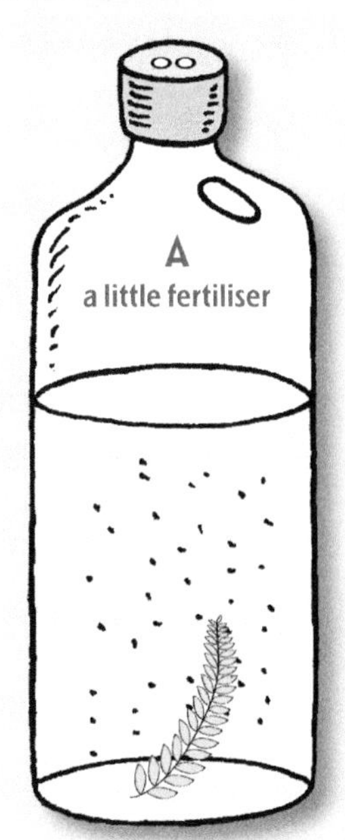

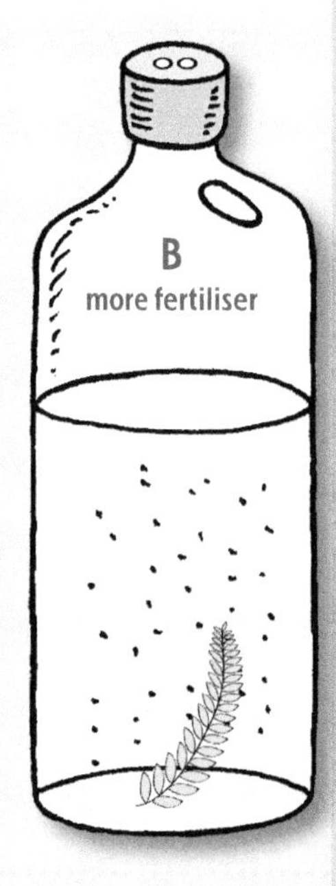

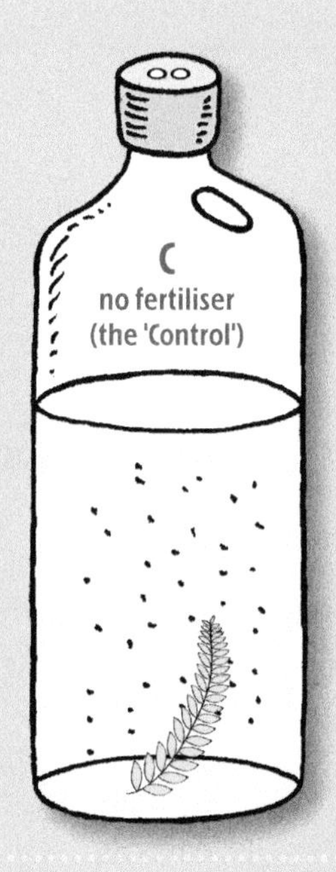

QUESTIONS

1. Name three chemical elements in fertiliser.
2. Explain how water gets from rivers up into clouds.
3. Why is the water cycle given the name 'cycle'?
4. State the meaning of each of these three words: tapu, tikanga, kaitiaki.
5. Explain what can happen if farmers stop using chemical fertilisers.
6. What does 'eutrophication' mean?
7. What is the main cause of eutrophication?
8. What do you think are the worst three aspects of eutrophication? Say why you chose these three.
9. How can fertiliser in a lake cause fish to die? Explain the connection.
10. Your body is 72% water. The water contained in your body this very moment was once in an iceberg and also in the Amazon jungle. Explain how this can be true.

Extension:

Find out what the NPK numbers on a fertiliser packet mean, and what other chemicals the packet also contains. Also, find where or how these chemicals are made. Explain how plants in nature get the chemicals they need without human help.

INDEX

ACKNOWLEDGEMENTS

iStock images – Cover, Vladimir Piskunov; page 5, Gerard Maas; page 6 (top) Johann Helgason, (bottom) Carmen Martínez; page 7, Barry Crossley; page 8 (diamond and wheel) Evgeny Terentyev; page 9 (archer) Darrell Fraser, (mouthguard) Maxim Borovkov; page 12 (water bottles) Doug Cannell, (kowhai) Sue McDonald; page 13 (fireworks) Stephen Strathdee, (spade) Bruce Lonngren; page 14 (fire) Shaun Lowe; (match) Serhiy Kobyakov; page 15, Alexey Tkachenko; page 16 (wave) Chuck Babbitt; page 19, Michael Walker; page 20 (heater) Sharon Meredith, (bulb) Gord Horne, (drill); page 21 (hair) Liza McCorkle, (lightning) Clint Spencer; page 23 (dam) Scott Espie, (solar panels) Peter Eckhardt; page 25 (solar panel) Mark Evans; page 26 (fox) Tara Minchin; page 28 (dog) Galina Barskaya, (magnet) Daniel Bobrowsky; page 30 Jennifer Altman; page 32 (elephant) Victor Soares, (cat) Mehmet Salih; page 33 (laptop) Nicholas Monu, (bbq) Anthony Clausen; page 34, Richard Goerg; page 35, Doug Nelson; page 36, Jan Rose; page 37, Pajara Thongjaj; page 41, Matthew Cole; page 42, Thomas Hottner; page 44, Jan Will; page 45, Leif Norman; page 50 (trumpeter) Rick Lord, (magnet) Thomas Mounsey; page 53 (fire) Greg Nicholas; page 54 (copper downpipe) Lew Zimmerman; page 55 (copper pipe) Konstantin Inozemtsev, (copper sulfate) Rainer Walter Schmied; page 57, Felix Möckel; page 59, Jennifer Trenchard; page 64 (river) Klaus Ferck; page 65 (marae) Greg Ward.
The didymo warning poster on page 66 is reproduced with permission of Biosecurity New Zealand.